Systems Biology 2014

Topical Conference at the 2014 AIChE Annual Meeting

Atlanta, Georgia, USA
16-21 November 2014

ISBN: 978-1-5108-1247-5

Printed from e-media with permission by:

Curran Associates, Inc.
57 Morehouse Lane
Red Hook, NY 12571

Some format issues inherent in the e-media version may also appear in this print version.

Printed by Curran Associates, Inc. (2015)

For permission requests, please contact AIChE
at the address below.

AIChE
120 Wall Street, FL 23
New York, NY 10005-4020

Phone: (800) 242-4363
Fax: (203) 775-5177

www.aiche.org

Additional copies of this publication are available from:

Curran Associates, Inc.
57 Morehouse Lane
Red Hook, NY 12571 USA
Phone: 845-758-0400
Fax: 845-758-2634
Email: curran@proceedings.com
Web: www.proceedings.com

TABLE OF CONTENTS

376855 Scalable, Accurate, and Systematic Engineering of Synthetic Pathways: Massively Multiplexed DNA Assembly, Next-Gen Sequence Verification, and Model-Guided Design

Abstract:

We have developed methods for DNA assembly, whole-construct sequencing, and (re-)design algorithms for large-scale optimization of multi-gene synthetic systems. In this work we applied these approaches to engineer a 16-gene fully synthetic and modular nitrogen-fixing device (100 parts, 23 kb). This synthetic nitrogen fixation (nif) gene cluster converts atmospheric nitrogen into ammonia, which is a critical process for global agricultural productivity. From characterized libraries of parts, we assembled >55,000 sequence-verified cistron modules for the nif gene cluster. This is a 100-1,000-fold improvement in library size over our and comparable previous studies. Approximately 10^{40} unique nif gene clusters can be built from this library. Additionally, we sequenced 7.5 million nif cistrons in total to quantify assembly error modes and frequencies. From these sequenced-verified cistrons, we used contact-free automated assembly to build permuted whole gene cluster designs. The gene cluster part substitutions and combinations were designed using algorithms based on design of experiments methodologies. In total we built, sequenced, tested, and modeled ~5 MB of systematically permuted gene cluster constructs using these approaches. We engineered improved nitrogen-fixing activity by tuning T7 promoter strength driving expression of each gene or operon flanked by strong terminators within the gene cluster. Multivariate statistical models were developed from these data for model-guided redesign and iterative engineering of the nif gene cluster. We believe the methods and strategies demonstrated here will accelerate mining and transferring biological functions between engineered systems and expand the complexity of synthetic systems that can be written, engineered, and optimized.

Decoders are combinational circuits that convert information from n inputs to a maximum of 2^n outputs. This operation is of major importance in computing systems yet it is vastly underexplored in synthetic biology. Here, we present a synthetic gene network architecture that operates as a biological decoder in human cells, converting 2 inputs to 4 outputs. As a proof-of-principle, we use small molecules to emulate the two inputs and fluorescent reporters as the corresponding four outputs. The experiments are performed using transient transfections in human kidney embryonic cells and the characterization by fluorescence microscopy and flow cytometry. We show a clear separation between the ON and OFF mean fluorescent intensity states. Additionally, we adopt the integrated mean fluorescence intensity for the characterization of the circuit and show that this metric is more robust to transfection conditions when compared to the mean fluorescent intensity. To conclude, we present the first implementation of a genetic decoder. This combinational system can be valuable toward engineering higher-order circuits as well as accommodate a multiplexed interface with endogenous cellular functions.

ACS Synthetic Biology

Research Article

pubs.acs.org/synthbio

Homology-Integrated CRISPR–Cas (HI-CRISPR) System for One-Step Multigene Disruption in *Saccharomyces cerevisiae*

Zehua Bao,[†,∥] Han Xiao,[‡,∥] Jing Liang,[‡] Lu Zhang,[‡] Xiong Xiong,[‡] Ning Sun,[†,⊥] Tong Si,[‡] and Huimin Zhao*[,†,‡,§]

[†]Department of Biochemistry, [‡]Department of Chemical and Biomolecular Engineering, [§]Departments of Chemistry, and Bioengineering, Institute for Genomic Biology, University of Illinois at Urbana–Champaign, Urbana, Illinois 61801, United States

Ⓢ *Supporting Information*

ABSTRACT: One-step multiple gene disruption in the model organism *Saccharomyces cerevisiae* is a highly useful tool for both basic and applied research, but it remains a challenge. Here, we report a rapid, efficient, and potentially scalable strategy based on the type II Clustered Regularly Interspaced Short Palindromic Repeats (CRISPR)–CRISPR associated proteins (Cas) system to generate multiple gene disruptions simultaneously in *S. cerevisiae*. A 100 bp dsDNA mutagenizing homologous recombination donor is inserted between two direct repeats for each target gene in a CRISPR array consisting of multiple donor and guide sequence pairs. An ultrahigh copy number plasmid carrying iCas9, a variant of wild-type Cas9, trans-encoded RNA (tracrRNA), and a homology-integrated crRNA cassette is designed to greatly increase the gene disruption efficiency. As proof of concept, three genes, *CAN1*, *ADE2*, and *LYP1*, were simultaneously disrupted in 4 days with an efficiency ranging from 27 to 87%. Another three genes involved in an artificial hydrocortisone biosynthetic pathway, *ATF2*, *GCY1*, and *YPR1*, were simultaneously disrupted in 6 days with 100% efficiency. This homology-integrated CRISPR (HI-CRISPR) strategy represents a powerful tool for creating yeast strains with multiple gene knockouts.

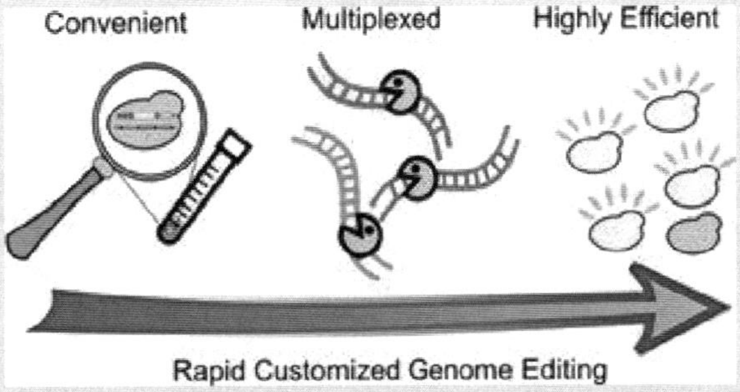

KEYWORDS: *CRISPR–Cas, Saccharomyces cerevisiae, multiple gene disruption, gene knockout, genome editing*

Multiple gene disruption is a widely used strategy for the characterization of specific gene functions within complicated genetic pathways.[1] It also facilitates the construction of versatile model hosts for heterologous gene expression,[2] identification of multidrug-resistant gene sets,[3] and investigation of disease mechanisms.[4] In addition, multiple gene disruption is also considered to be an effective strategy for improving biofuels and biochemicals production in metabolic engineering.[5,6]

As a model organism, *Saccharomyces cerevisiae* has been widely used for genetic research,[7] exploration of disease mechanisms,[8] and metabolic engineering.[9] Single gene disruption in *S. cerevisiae* is typically achieved by substituting the target gene with a selectable marker through homologous recombination (HR), which is highly efficient in *S. cerevisiae*.[10] Therefore, multiple gene disruption is traditionally achieved in a sequential manner,[11] by recycling the limited number of selectable markers. Unfortunately, challenges inevitably arise during this process. Not only is this process slow and tedious, but also marker removal can frequently cause mistargeting events.[12,13] Very few strategies exist today to perform simultaneous multiple gene disruption in yeast. A recently developed tool, Green Monster, allows deletion of multiple genes in *S. cerevisiae* via cycles of sexual assortment and fluorescence activated cell sorting (FACS).[14] However, each deletion target has to be first replaced by a reporter gene, and iterative rounds of mating and sporulation are required, which is very laborious and time-consuming.[14]

The type II bacterial Clustered Regularly Interspaced Short Palindromic Repeats and CRISPR-associated proteins (CRISPR–Cas) system was recently exploited as an efficient gene-targeting technology in several prokaryotes and eukaryotes.[15−18] A trans-activating crRNA (tracrRNA):crRNA duplex (or a chimeric guide RNA (gRNA)) directs the Cas9 protein to cleave a target DNA sequence with a required protospacer adjacent motif (PAM), which is composed of any of the four bases followed by two guanines.[19] Simultaneous delivery of multiple gRNAs opens up the possibility for targeting multiple genes at the same time.[15] In a recent work, a near 100% single gene replacement efficiency was reached after cotransformation of the CRISPR–Cas system and a linearized HR donor DNA in *S. cerevisiae*.[20] However, there are several limitations for multiple gene targeting using this strategy. First, either 5 μg of marker cassette (1.4 kb) or 1 nmol (~58.5 μg) of double-stranded oligonucleotide donor (90 bp) is required to achieve a 100% single gene replacement efficiency.[20] It is hard to ensure that this amount of DNA will all be accepted by *S.*

Received: May 28, 2014

ACS Publications

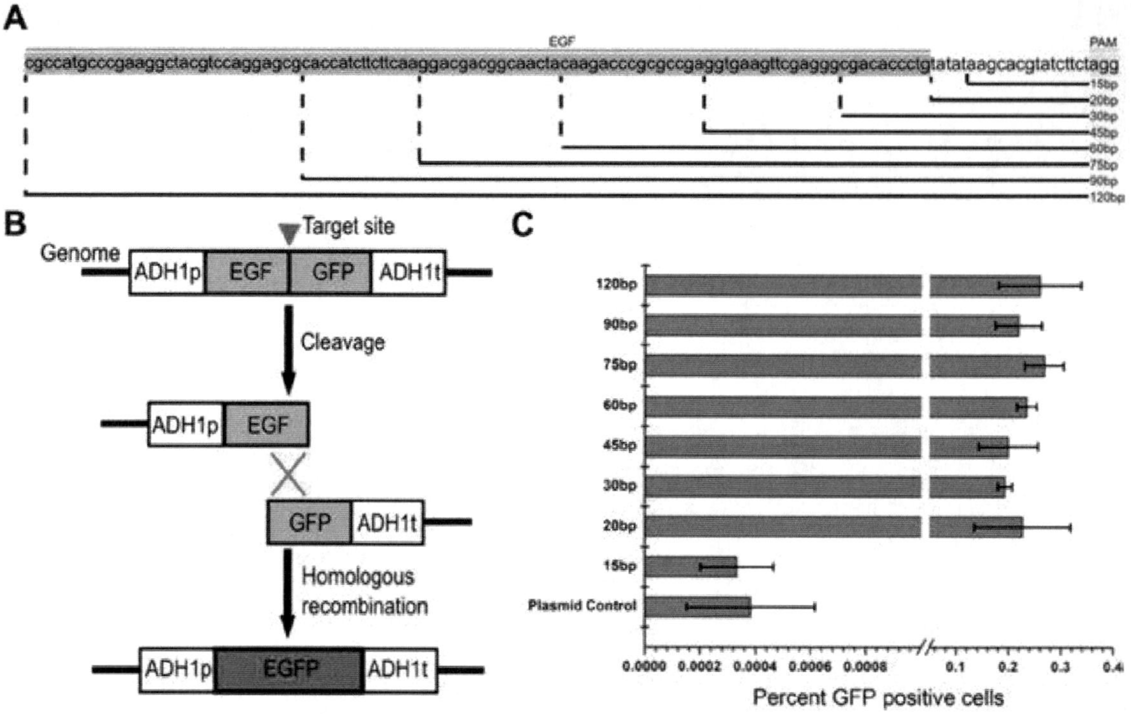

Figure 1. CRISPR–Cas targeting EGFP reporter with guide sequence GFP0. (A) Scheme showing the design of GFP0 spacers with different lengths. The PAM sequence is highlighted in yellow. The 3′ end of EGF sequence is highlighted in gray. (B) Scheme showing the design of the EGFP reporter system. EGFP function will be restored after Cas9 cleavage and homologous recombination. (C) Percentage of GFP positive cells 48 h after transformation of the CRISPR–Cas system with different GFP0 spacers. Error bars indicate standard deviation of three biological repeats.

cerevisiae when multiple donors need to be introduced.[21,22] Second, as observed with other genome editing tools (e.g., transcription activator-like effector nucleases (TALENs)),[23,24] the targeting efficiency mediated by CRISPR–Cas is also site-dependent.[25,26] The 100% gene replacement efficiency has been reported only for targeting CRISPR-susceptible sites.[20] Third, several RNA polymerase III regulatory elements are needed to express gRNAs targeting multiple genes,[18,19] which is inconvenient for standardized plasmid construction. Therefore, it is not surprising that this strategy has not been used to generate multiple gene disruptions in *S. cerevisiae*.

Using the CRIPSR–Cas from *Streptococcus pyogenes*,[27] we developed a HI-CRIPSR system to address these limitations. To the best of our knowledge, this is the first example of using CRISPR–Cas for multiple gene disruption in *S. cerevisiae*. As proof of concept, three endogenous genes, *CAN1*, *LYP1*, and *ADE2*, were simultaneously disrupted. The overall efficiency ranged from 27 to 87% when different sets of targeting sites were used. Furthermore, another three genes involved in an artificial hydrocortisone biosynthetic pathway, *ATF2*, *GCY1*, and *YPR1*, were simultaneously disrupted in 6 days with 100% efficiency, which was determined by random genotyping.

■ RESULTS AND DISCUSSION

Identification of the Minimum Length of a Targeting Sequence. During the process of crRNA maturation, the 5′-end region of the spacer is cleaved by RNase III and unknown endogenous nuclease(s), which means that this region does not contribute to crRNA function.[27,28] Therefore, we sought to exploit this region to harbor HR disruption donors in such a way that multiple crRNAs and HR disruption donors can be assembled in one step. To this end, the minimal length of a functional spacer was identified using an enhanced green fluorescent protein (EGFP)-based single-strand annealing reporter system (Figure 1).[29] Repair of the two truncated EGFP DNA fragments can be triggered by a double strand break (DSB) generated by CRISPR–Cas, resulting in the expression of a functional EGFP (Figure 1B). A spacer with 20 bp upstream of the PAM sequence successfully restored the function of EGFP, and comparable expression of functional EGFP was also observed when the spacer length was increased to 120 bp (Figure 1C).

Design of the HI-CRISPR System. The core design principle of the HI-CRISPR system is to achieve efficient gene disruption in *S. cerevisiae* with simple manipulations. The HR disruption donor containing an 8 bp deletion in the middle is designed to recombine with the target genomic locus (Figure 2A). The 8 bp deletion includes the PAM sequence and the last 3 bp of the protospacer, which prevents continuous recognition and cleavage by CRISPR–Cas after recombination. To avoid repeated introduction of dsDNA fragments as repairing templates, the HR disruption donor was maintained under selective pressure by integrating the donor between two direct repeats as a 5′ extension of the guide sequence (Figure 2B), which enables one-step construction of all required elements (Figure S1). In the CRISPR array, expression of multiple spacers flanked by direct repeats is driven by a single promoter, where the resultant transcript can be processed into multiple crRNAs.[30] To take advantage of natural polycistronic CRISPR arrays for multiplexed targeting, crRNA is used instead of gRNA (Figure 3A). The transcribed pre-crRNA can be processed into mature crRNA by host RNase III and unknown

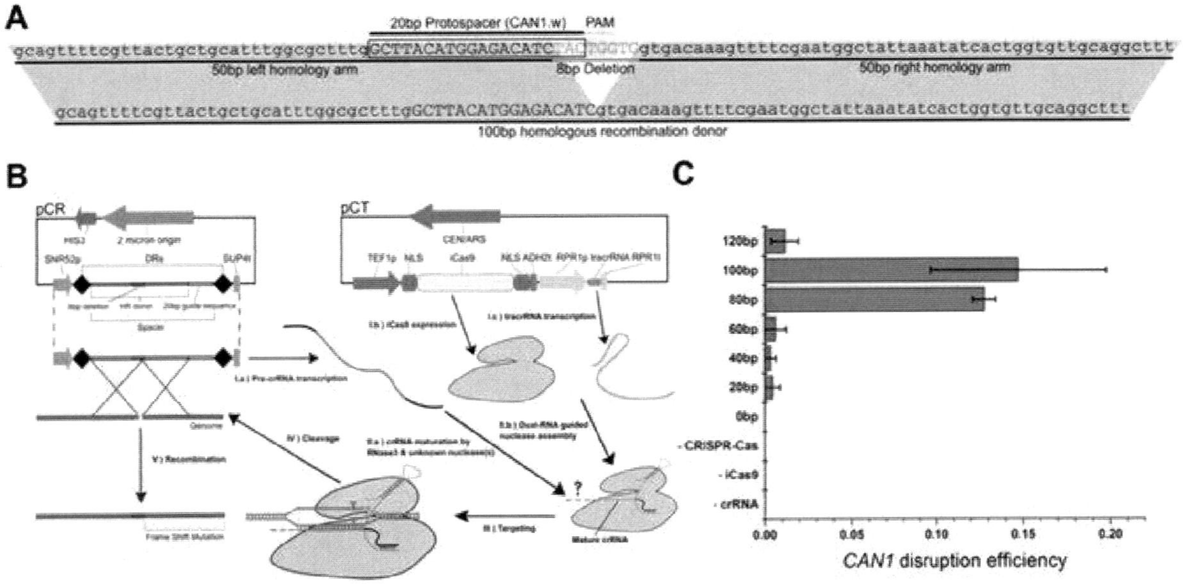

Figure 2. Determination of the optimal HR donor length for efficient single gene disruption. (A) Scheme showing the design of a 100 bp HR donor for the CAN1.w site. The 8 bp deletion includes the PAM sequence and the last 3 bp of the guide sequence. HR donors with different lengths were obtained by simultaneously changing the 5′ end of left homology arm and the 3′ end of right homology arm by 10 bp. (B) Schematic of the two-plasmid HI-CRISPR system for single gene disruption. The crRNA array was constructed using the pRS423 backbone to generate plasmid pCR. iCas9 and tracrRNA were constructed using the pRS415 backbone to generate plasmid pCT. The pre-crRNA is transcribed from the pCR plasmid and processed into mature crRNA by RNase III and unknown nuclease(s). The green dashes at the 5′ end of mature crRNA and the question mark indicate that it is unknown whether the HR donor is fully cleaved. iCas9 and tracrRNA are expressed from the pCT plasmid and complexed with mature crRNA to form the dual-RNA-guided nuclease. The nuclease is then guided to target the genomic locus. The target gene is cleaved by iCas9, generating a DSB. The DSB is repaired through recombination with the HR disruption donor harbored in the spacer of the crRNA array on plasmid pCR. The repaired gene function is disrupted by a small deletion in the donor sequence. (C) *CAN1* disruption efficiency obtained 2 days after transformation. Error bars indicate standard deviation of two biological repeats.

nuclease(s). Cas9 and tracrRNA are expressed separately and then assembled with crRNA to form the dual-RNA-guided endonuclease. The nuclease is then guided by crRNA to target and cleave the genomic locus to generate a DSB. The DSB is repaired through recombination with the HR disruption donor. Once the HR disruption donor is recombined with the target locus, a frame shift will be generated to abolish the normal protein translation (Figure 2B). In most cases, a premature stop codon was consequently introduced.

Discovery of a Cas9 Variant. A Cas9 variant was accidentally obtained during the plasmid construction process. This gene harbors two point mutations (Figure S2), resulting in a double mutant variant (D147Y, P411T). Gene disruption efficiency mediated by the variant was higher than that of wild-type Cas9 (Figure S3). This gene was named iCas9 (improved Cas9) and therefore adopted for all subsequent experiments.

Determination of the Optimal Length of an HR Disruption Donor. The optimal length of an HR disruption donor contained in a spacer was further determined for single gene disruption by CRISPR–Cas. The endogenous negative selectable gene *CAN1*, encoding plasma membrane arginine permease, was chosen as the target. The disruption mutant of *CAN1* grew in the presence of toxic arginine analogue canavanine, enabling a quick readout of disruption efficiency.[31] HR disruption donors with 20 bp increments were designed by simultaneously changing the 5′ end of the left homology arm and the 3′ end of the right homology arm by 10 bp (Figure 2A). A two-plasmid system was used for CRISPR–Cas expression (Figure 2B). The iCas9 gene and tracrRNA, driven by the TEF1p and RPR1p promoters, respectively, were cloned into the centromeric plasmid pCT (Table S1). Together with the 20 bp guide sequence targeting 1331 bp downstream of the start codon of *CAN1* (site CAN1.w), the 20, 40, 60, 80, 100, and 120 bp HR disruption donors were constructed on the 2 μ plasmid pCR (Table S1) for *CAN1* disruption, achieving gene disruption efficiencies of 0.44, 0.33, 0.62, 12.76, 14.71, and 1.16% on day 2, respectively (Figure 2C). *CAN1* gene disruption was not detectable in the absence of an HR disruption donor and a crRNA (Figure 2C). Similarly, no gene disruption was detectable when only the HR donor and crRNA were present (Figure 2C). We sequenced 16 randomly selected canavanine-resistant colonies and verified the expected 8 bp deletion at the targeted locus in all clones (Figure S4), indicating that the loss of *CAN1* function was caused by CRISPR–Cas assisted homologous recombination.

The highest *CAN1* disruption efficiency (14.71%) was obtained using a 100 bp HR disruption donor. Therefore, this donor length was chosen for subsequent gene disruptions. However, this efficiency was not yet high enough for multiplex gene targeting. We speculate that suboptimal cleavage or low HR efficiency can be improved by giving the cells more time to grow. Thus, we sought to improve the gene disruption efficiency by growing the cells for a prolonged period of time in liquid SC medium, which prevented the loss of CRISPR–Cas plasmids. As shown in Figure S5, *CAN1* disruption efficiency was steadily improved through serial transfers in liquid SC-LH medium. After 12 days of culture in liquid medium, a 100% *CAN1* disruption efficiency was obtained.

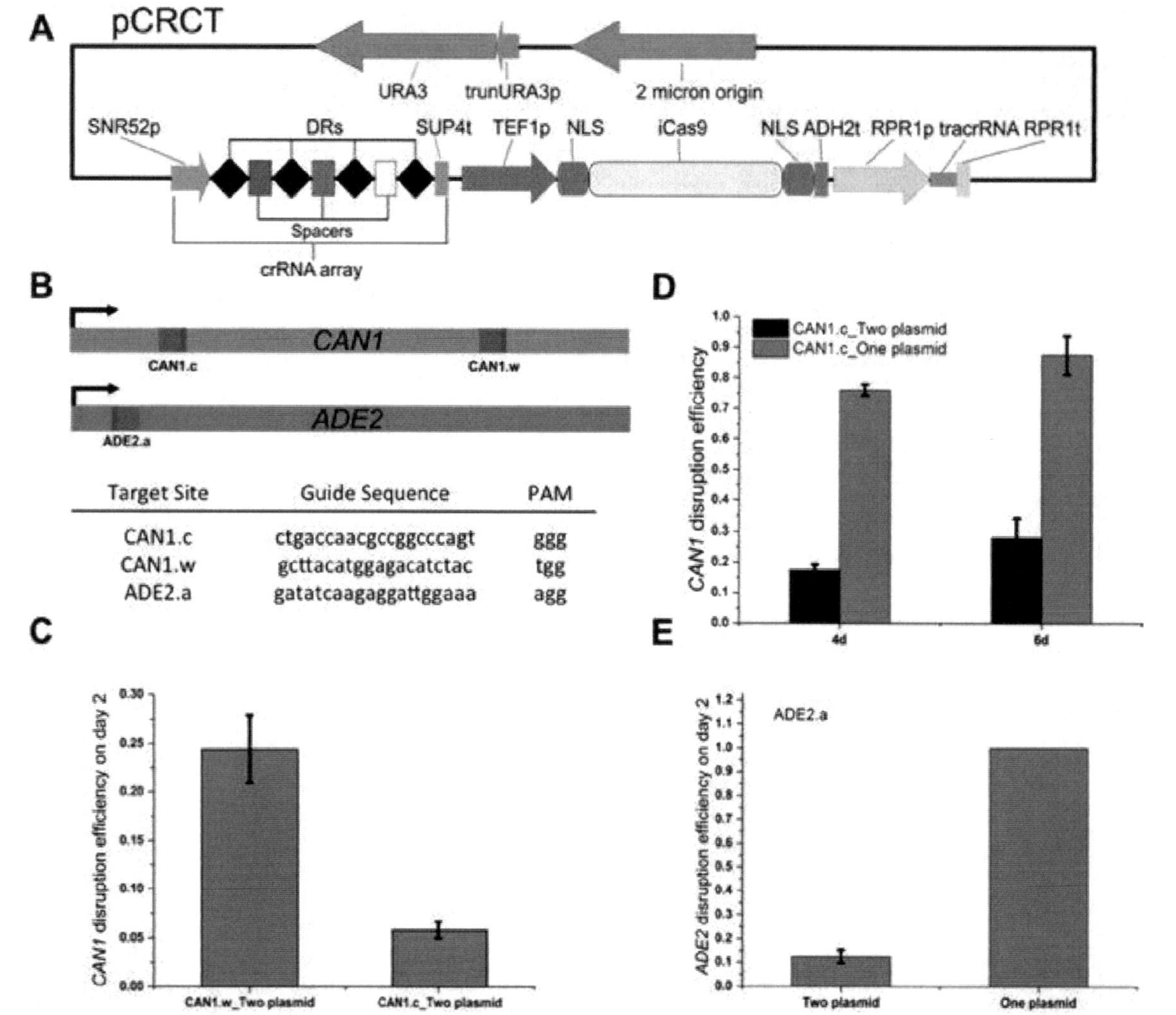

Figure 3. Gene disruption efficiency by the one-plasmid HI-CRISPR system. (A) Scheme showing the design of one-plasmid HI-CRISPR system. (B) The upper scheme shows the relative positions of the CAN1.w, CAN1.c, and ADE2.a target sites within each gene. The gray box represents the guide sequence, and the purple bar represents the PAM sequence. A PAM sequence on the left indicates a site on the negative DNA strand. The table summarizes the guide sequence and PAM sequence of each target site. (C) *CAN1* disruption efficiencies were assessed 2 days after transformation of the two-plasmid system targeting either CAN1.w or CAN1.c. (D) *CAN1* disruption efficiencies were assessed 4 or 6 days after transformation of either two-plasmid or one-plasmid CRISPR–Cas systems targeting the CAN1.c site. (E) *ADE2* disruption efficiency was assessed 2 days after transformation of the one-plasmid system targeting at the ADE2.a site. All error bars indicate standard deviations of three biological repeats.

Stimulation of Gene Disruption by Increasing the Plasmid Copy Number. Although a high efficiency of single gene disruption was achieved, it was time-consuming. In order to shorten the time needed for efficient gene disruption, we attempted to stimulate gene disruption by increasing the copy number of the plasmid harboring the HI-CRISPR system. Previous studies showed that the *URA3* gene with a truncated promoter on a plasmid was poorly expressed in yeast cells. To survive under the selection pressure, the cell needs to maintain a much higher copy number (~200 copies per cell) than that of a typical 2 μ-based plasmid (10–40 copies per cell).[32,33] Therefore, the ultrahigh copy number receiver plasmid pCRCT was constructed for expressing the CRISPR–Cas gene and harboring the HR disruption donor (Figure 3A). As expected, a 100% *CAN1* disruption efficiency was obtained after only 4 days of cultivation in selective medium when the CAN1.w site was targeted. To further demonstrate the efficacy of this one-plasmid-based HI-CRISPR system, a poorly targeted site of the *CAN1* gene (site CAN1.c), 362 bp downstream of the start codon (Figure 3B), was also chosen for analysis. Under the two-plasmid system, the *CAN1* disruption efficiency with the CAN1.c target was 5.82% after 2 days of cultivation in liquid culture, which was 76% lower than that with the CAN1.w target (Figure 3C). Under the single plasmid system, the *CAN1* disruption efficiencies were 76 and 87% after 4 and 6 days of cultivation in liquid medium, respectively. In comparison, the corresponding disruption efficiencies of the CAN1.c target were only 18 and 28%, respectively, using the two-plasmid system (Figure 3D).

To investigate whether the improved gene disruption efficiency was generally applicable, disruption of another gene, *ADE2*, was performed. The *ADE2* gene encodes a

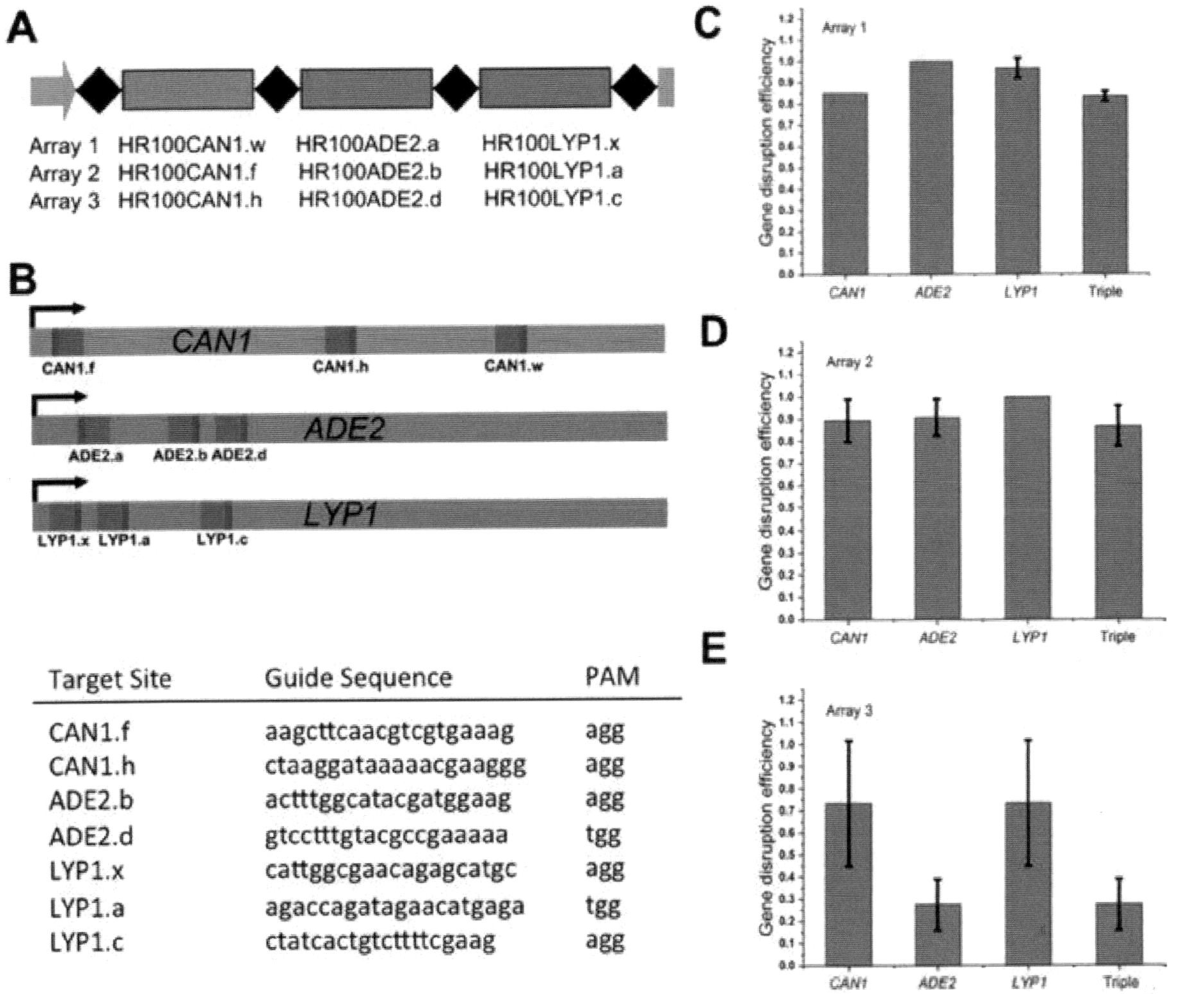

Target Site	Guide Sequence	PAM
CAN1.f	aagcttcaacgtcgtgaaag	agg
CAN1.h	ctaaggataaaaacgaaggg	agg
ADE2.b	actttggcatacgatggaag	agg
ADE2.d	gtcctttgtacgccgaaaaa	tgg
LYP1.x	cattggcgaacagagcatgc	agg
LYP1.a	agaccagatagaacatgaga	tgg
LYP1.c	ctatcactgtcttttcgaag	agg

Figure 4. Simultaneous triple gene disruption using the one-plasmid HI-CRISPR system. (A) Schematic illustration of three different HI-CRISPR arrays. (B) Scheme showing the relative positions of target sites within each gene used in triple gene disruption experiments. The guide and PAM sequences of each site are summarized in the table. (C–E) Single gene and triple gene disruption efficiencies assessed 4 days after transformation of the corresponding CRISPR arrays using the one-plasmid system. Error bars indicate standard deviation of three biological repeats.

phosphoribosylaminoimidazole carboxylase required for adenine biosynthesis. Red pigment will accumulate in the *ADE2* nonsense mutant cell when adenine is absent from the medium.[34] When targeting 154 bp downstream of the start codon of the *ADE2* gene (site ADE2.a, Figure 3B), a 100% disruption efficiency was achieved with the plasmid pCRCT-ADE2a after 2 days in liquid culture, whereas only a 12.5% *ADE2* disruption efficiency was obtained with the two-plasmid system consisting of pCT and pCR-ADE2a at the same time point (Figure 3E). Taken together, these results demonstrated that increasing the plasmid copy number was an effective strategy for improving gene disruption efficiency mediated by CRISPR–Cas.

One-Step Multiple Gene Disruption by the HI-CRISPR System. Next, multiple simultaneous gene disruptions mediated by the HI-CRISPR system was investigated. In addition to *CAN1* and *ADE2*, the *LYP1* gene, encoding lysine permease, was chosen as the third target for multiple disruption. Similar to that for the *CAN1* disruption mutant, the *LYP1* disruption mutant can grow in the presence of thialysine, a toxic lysine analogue.[35] Three sites, including CAN1.w (1331 bp downstream of the start codon of *CAN1*), ADE2.a (154 bp downstream of the start codon of *ADE2*), and LYP1.x (79 bp downstream of the start codon of *LYP1*), which showed a high efficiency for corresponding single gene disruption (Figures 3 and S6), were combined together as array 1 for multiple gene disruption (Figure 4A). After 4 days of culture in SC-U medium, an 83% triple gene disruption efficiency was achieved using array 1 (Figure 4C and Table S2).

To evaluate the general applicability of this HI-CRISPR system for simultaneous multigene disruption, another two CRISPR arrays were designed for these three target genes without prescreening the target sites. To minimize the possibility of choosing a low-efficiency target site and to ensure the targeting specificity of the CRISPR–Cas system, several criteria were adopted: (i) An earlier region of the gene is preferred for targeting to ensure disruption of gene function. (ii) According to a previous study, one base pair mismatch at the last 12 bp of the guide sequence abolished the cleavage activity of CRISPR–Cas.[18] Therefore, the last 12 bp of the

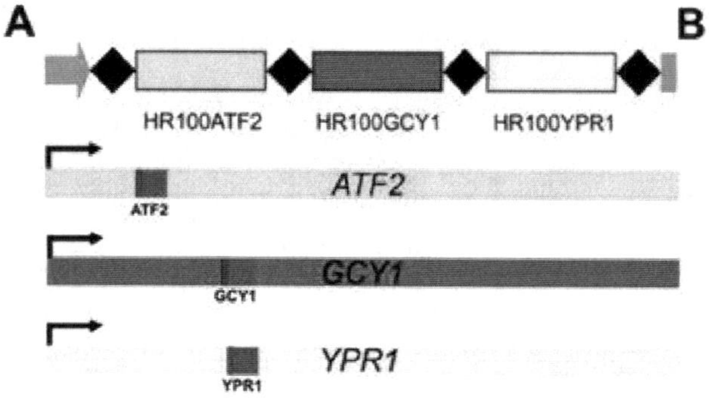

Target Site	Guide Sequence	PAM
ATF2	atgaaatcatgctgtggaaa	agg
GCY1	tcggtagccatctttcaaag	cgg
YPR1	gccgaaacccaacactggaa	tgg

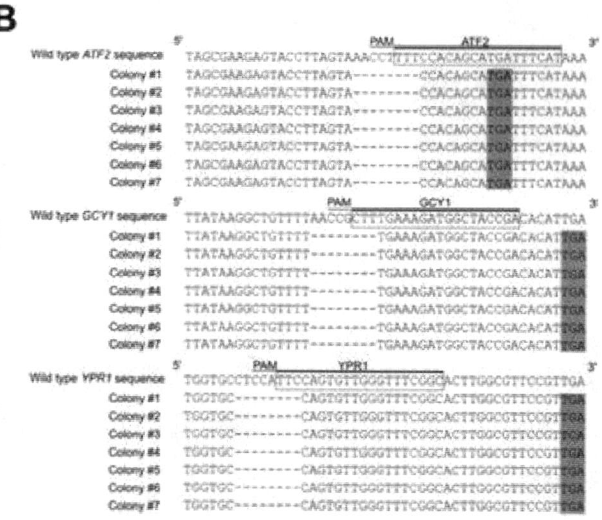

Figure 5. Simultaneous disruption of three genes involved in a hydrocortisone biosynthetic pathway using the one-plasmid HI-CRISPR system. (A) Schematic of the HI-CRISPR array and relative positions of target sites within each gene. The guide and PAM sequences of each site are summarized in the table. (B) Alignment of gene sequences from seven randomly picked colonies against the wild-type sequences. All seven colonies were triple disruption mutants.

guide sequence should be unique in the genome of *S. cerevisiae*, minimizing off-target activity. On the other hand, a genetic screen study using CRISPR−Cas showed that the gRNA efficiency is associated with specific sequence characteristics.[26] These specific characteristics are also included in the design criteria: (iii) purines are preferred in the last 4 bp of spacer, (iv) the antisense strand is preferred for targeting, and (v) the GC content of the 20 bp guide sequence should be between 20 and 80%. In addition, type III RNA polymerase terminates transcription at a small polyT stretch.[36] So, (vi) polyT (more than 4 T's) sequences are avoided in both the HR disruption donor and the crRNA guide sequence.

As a result, array 2 was designed to target 65 bp downstream of the start codon of *CAN1* (site CAN1.f), 473 bp downstream of the start codon of *ADE2* (site ADE2.b), and 112 bp downstream of the start codon of *LYP1* (site LYP1.a). Using array 2 for multiple gene targeting, a triple gene disruption efficiency (87%) was achieved in 4 days, similar to that using array 1 (Figure 4C−D and Table S2). Array 3 was designed to target 846 bp downstream of the start codon of *CAN1* (site CAN1.h), 555 bp downstream of the start codon of *ADE2* (site ADE2.d), and 543 bp downstream of the start codon of *LYP1* (site LYP1.c). After 4 days culture in SC-U medium, the triple gene disruption efficiency reached 27% (Figure 4E and Table S2).

To demonstrate that the HI-CRISPR system can be used to disrupt genes whose mutations do not lead to clear phenotypes, another three yeast endogenous genes involved in an artificial hydrocortisone biosynthetic pathway, *ATF2*, *GCY1*, and *YPR1*,[37] were chosen as candidates for disruption. Three target sites, ATF2, GCY1, and YPR1, were designed according to the criteria and assembled together with corresponding HR disruption donors (Table S4) into a single crRNA array (Figure 5A). After 6 days of culture in SC-U medium, single colonies were isolated by diluted plating, and seven of them were randomly picked and screened directly by PCR and sequencing. All seven colonies were shown to be triple disruption mutants at the defined target sites (Figure 5B).

Investigation of Increased Multiplexing. To investigate how many functional crRNAs can be contained in such a CRISPR array, nine crRNAs with only one functional crRNA targeting the CAN1.w site of the *CAN1* gene in different positions of the CRISPR array were designed (Figure 6A). When the functional crRNA was placed at the first position, the *CAN1* disruption efficiency was similar to that observed using only one crRNA (Figure 6B). However, the *CAN1* disruption

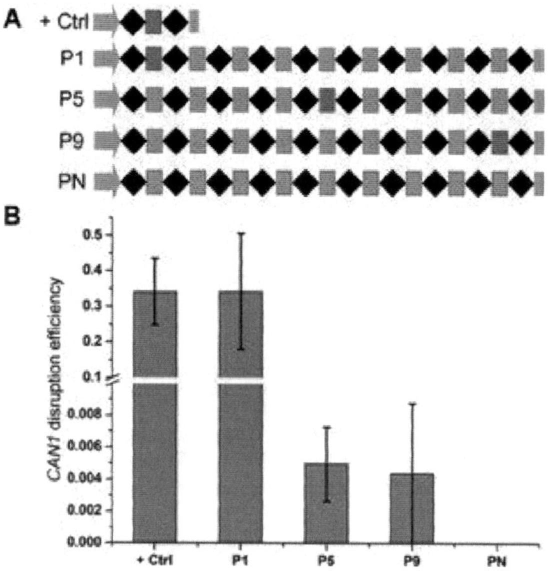

Figure 6. Investigation of the crRNA number tolerance for use in a single crRNA array. (A) Scheme showing the design of P1, P5, P9, and PN crRNA arrays. The red box represents the genome targeting HR80CAN1.w spacer, whereas the gray boxes represent nontargeting spacers with random sequences. (B) *CAN1* disruption efficiencies calculated 6 days after transformation. Error bars indicate standard deviations of two biological repeats.

Table 1. Comparisons among Different Multiple Simultaneous Gene Disruption Techniques in *S. cerevisiae*

methods	Cre−LoxP[13]	Green Monster[14]	this work
genes disrupted	4	16	3
time spent	35 days	143 days	4 days
average time per gene	8.75 days	8.94 days	1.33 days
selection or screening needed	auxotrophic marker selection	FACS	no
labor	intensive	intensive	low
ectopic sequences introduced	yes	yes	no
plasmid construction	multiple PCR	multiple PCR	one step Golden Gate
other constraints	selectable markers limited	(1) Single gene disruption mutant needed; (2) genotyping after each round; (3) some genes cannot be targeted.	target site dependency

efficiency dropped to near zero when this crRNA was placed at the fifth position (Figure 6B).

By taking advantage of the crRNA maturation mechanism of CRISPR−Cas and physiological properties of *S. cerevisiae*, we developed a new strategy for efficient and user-friendly multiple simultaneous gene disruption in *S. cerevisiae*. Three features distinguish this system from other related methods: (i) customized crRNAs containing HR disruption donors allow easy construction in one step, (ii) gene disruption efficiency was greatly improved using an ultrahigh copy number plasmid, and (iii) multiple gene disruption mutants can be easily obtained by random genotyping without screening or selection. Although only three genes were efficiently disrupted by this system, the plasmid can be removed by counter selection of *URA3* and recycled for the next round of multiple gene disruption. Therefore, this strategy is potentially scalable. As summarized in Table 1, compared to other existing methods, this strategy is more efficient, less time-consuming, and easier to manipulate. However, this strategy also has limitations. It was noted that high gene disruption efficiency was attained after a certain time period of incubation in liquid medium, making the isolation of gene disruption mutants with growth defects problematic.

An interesting finding of this study was iCas9. The two mutations (D147Y, P411T) were in the REC1 domain of Cas9, which is a critical region for DNA cleavage.[38] This may contribute to the improved targeting efficiency by iCas9. Preliminary studies also indicated that no obvious off-target effect was found when using HI-CRISPR for multiple gene disruption (data not shown), suggesting the potential broad applicability of iCas9 for genome editing.

The *CAN1* disruption efficiency with a 120 bp HR disruption donor was much lower than that with a 100 bp HR disruption donor (Figure 2C), which is unexpected. The most likely cause was that the 120 bp HR donors gave rise to a deficient CRISPR−Cas system. In fact, in contrast to the 100 bp HR donor, a poly-T sequence was found in the 120 bp HR donor, which might cause premature transcriptional termination of the crRNA. In addition, it has been reported that the sequence downstream of the promoter, recognized by type III RNA polymerase, affects the promoter strength.[39] Therefore, the sequence of the 120 bp HR donor might also account for the decreased transcriptional strength of crRNA.

The all-in-one HI-CRISPR system using a single ultrahigh copy number plasmid greatly improved the gene disruption efficiency (Figure 3). Compared to nonhomologous end joining, *S. cerevisiae* always prefers HR to repair a DSB.[40] The overall efficiency of CRISPR−Cas mediated gene disruption in *S. cerevisiae* therefore depends on two aspects: the efficiency of DSB introduction by CRISPR−Cas and the efficiency of the subsequent HR repair. The use of an ultrahigh copy number plasmid may have contributed to both processes. First, increasing the copy number of the plasmid is considered to be an effective strategy for increasing enzyme expression in *S. cerevisiae*,[41] which, in our case, may have helped with Cas9 expression and led to higher cleavage efficiency. Second, under the all-in-one system, increasing the copy number of the plasmid also increases the availability of the HR disruption donor. Since we have already optimized the length of the donor based on *CAN1* disruption efficiency using the two-plasmid system (Figure 2C), increasing donor availability may further improve HR efficiency.

Different triple gene disruption efficiencies were observed using different CRIPSR arrays (Figure 4), indicating that the site dependence of CRISPR−Cas is still a challenge for achieving a higher efficiency of multiple simultaneous gene disruption. Although this effect is partially alleviated through the use of the ultrahigh copy number plasmid pCRCT (Figure 3), better rules for choosing efficient CRISPR targeting sites are still needed to obtain multiple gene disruption mutants more reliably. Alternatively, using multiple crRNAs to target each gene may address the site dependence issue. However, our results suggest that the maximum number of functional crRNAs will be no more than four (Figure 6). Possible reasons for this are the incomplete transcription and/or instability of the long CRISPR RNA. Developing regulation elements to enhance the expression and/or improve the stability of CRISPR RNA may address this limitation.

In addition to multiple simultaneous gene disruption, there are other potential applications of this HI-CRISPR system. A comparable gene disruption efficiency was found between the 100 bp and 80 bp HR disruption donors (Figure 2C). Taking advantage of the spared length, an HR disruption donor can be designed to make a modification instead of deletion of the target locus with the assistance of CRISPR−Cas (e.g., point mutations and tag insertions). Moreover, the 120 bp homology-integrated spacer was the only difference among every single gene disruption cassette. Such length of oligos can be synthesized on a CustomArray (CustomArray Inc.) in a complex pool,[26] and a library of gene modification cassettes can be constructed using this system. Given its flexibility and potential applications, this HI-CRISPR system can be a valuable tool to conduct genome scale studies in *S. cerevisiae*.

■ METHODS

Strains and Media. The EGFP reporter strain was constructed elsewhere based on the *S. cerevisiae* strain HZ848 (MATα *ade2-1 ade3*Δ*22 ura3*Δ*0 his3-11,15 trp1-1 leu2-3,112* *and can1-100*).[29] The *S. cerevisiae* strain BY4741 (MATa *his3Δ0 leu2Δ0 met15Δ0 ura3Δ0*) was used in this study for assessing *CAN1*, *ADE2*, and *LYP1* gene disruption efficiency. The *S. cerevisiae* strain CEN.PK2-1c (MATa *ura3-52 trp1-289 leu2-3,112 his3Δ1 MAL2-8^C SUC2*) was used for disruption of *ATF2*, *GCY1*, and *YPR1*. YPAD media was used to grow yeast cells before transformation. After transformation, cells were grown in appropriate synthetic complete (SC) media minus the auxotrophic compound complemented by the plasmids. SC-LHR (SC-leucine, histidine, and arginine) and SC-UR (SC-uracil and arginine) agar plates with 60 μg/mL L-canavanine (Sigma) were used to select for *CAN1* disrupted cells. SC-LHK (SC-leucine, histidine, and lysine) and SC-UK (SC-uracil and lysine) agar plates with 250 μg/mL thialysine (S-2-aminoethyl-L-cysteine, Sigma) were used to select for *LYP1* disrupted cells. SC-LH (SC-leucine and histidine) and SC-U (SC-uracil) agar plates minus adenine hemisulfate were used to screen for *ADE2* disrupted cells.

CRISPR–Cas Target Site Selection and Donor Design. A 12 bp seed sequence together with NGG PAM sequence (N_{12}NGG) was searched on both strands of the target gene sequence and blasted against the *S. cerevisiae* S288c genome (NCBI Taxonomy ID: 559292). The unique sequences were selected as target sites to minimize off-target effects.[18] A 100bp donor sequence was designed to have two 50 bp homology arms flanking the Cas9 cutting site and incorporate an 8 bp deletion including the PAM sequence, thus introducing a frame-shift mutation (Figure 2A,B). All donor and guide sequences were examined and did not contain more than 4 T residues in a row to prevent early termination of RNA transcription.

Plasmid Construction. The RPR1p, RPR1t, and SNR52p regulatory elements were PCR-amplified from wild-type BY4741 genomic DNA. SUP4t was directly synthesized into cloning primers. The CRISPR RNA elements DR and tracrRNA were amplified from plasmid pX260, which was a gift from Dr. Feng Zhang.[18] The human codon-optimized Cas9 gene used in the EGFP reporter assay was also amplified from plasmid pX260. The Cas9 gene in pGal10p-Cas9 was originally from *S. pyogenes* SF370 and PCR amplified from plasmid pMJ806, which was a gift from Dr. Jennifer A. Doudna.[42] This gene was tagged with a FLAG tag and an SV40 nuclear localization signal on its N-terminus as well as an SV40 nuclear localization signal on its C-terminus. The two point mutations in iCas9 sequence were accidentally introduced during construction of plasmid pCT (Figures S2 and S7 and Tables S1 and S3). All of the spacer sequences were ordered as gBlocks from Integrated DNA Technologies (Coralville, IA) and assembled into pCR or pCRCT plasmids using the Golden Gate assembly method (Figure S1).[43] All of the sequence information, primers used for plasmid construction, and the iCas9 sequence can be found in Supporting Information (Tables S4–S8 and Figure S2). The plasmid descriptions can be found in Supporting Information Table S1. The plasmid construction processes are summarized in Supporting Information Table S3 and Figures S1 and S7.

Yeast Transformation. Plasmid transformation of BY4741 and CEN.PK2-1c (1 μg of each plasmid per transformation) was carried out using LiAc/SS carrier DNA/PEG method.[44] After transformation, cells were recovered in 1 mL of YPAD at 250 rpm and 30 °C for 1 h, washed with water once, transferred to 2 mL of the appropriate SC media, and cultivated at 250 rpm, 30 °C. Every 48 h, 100 μL of cell culture was transferred into 2 mL of fresh SC media to ensure cell viability. The remaining cell culture was discarded.

Calculation of Single Gene Disruption Efficiency. To determine the optimal HR donor length, *S. cerevisiae* BY4741 cells were transformed with pCT and pCR-CAN1.w with different HR donor lengths. Two-hundred microliters of 10^4-fold diluted cell culture was plated onto selective and nonselective plates on the appropriate days. Cells were allowed to grow for 2 days on the plate, and the *CAN1* single gene disruption efficiency was determined by calculating the ratio of yeast colonies formed on selective and nonselective plates. Experiments were done in duplicate.

To compare the two-plasmid and one-plasmid CRISPR–Cas systems, cells were transformed with either the two-plasmid or one-plasmid CRISPR–Cas system targeting different sites. Two-hundred microliters of 10^4-fold diluted cell culture was plated onto selective and nonselective plates on the appropriate days. Cells were allowed to grow for 2 days, and the *CAN1* single gene disruption efficiency was determined by calculating the ratio of yeast colonies formed on selective and nonselective plates. The *ADE2* single gene disruption efficiency was determined by calculating the percentage of pink colonies on SC plates minus adenine hemisulfate. Experiments were done in triplicate.

Calculation of the Triple Gene Disruption Efficiency. *S. cerevisiae* BY4741 cells were transformed with the one-plasmid CRISPR–Cas system targeting the *CAN1*, *ADE2*, and *LYP1* genes. Two-hundred microliters of 10^4-fold diluted cell culture was plated onto SC-U plates on day 4. After 2 days, a total of 20 (array 1) or 50 (arrays 2 and 3) colonies were randomly selected and streaked onto selective plates. After another 2 days, growth and color of each colony were recorded. The triple gene disruption efficiency was determined by calculating the proportion of colonies with all three genes disrupted. Experiments were done in triplicate.

Yeast Genomic DNA Extraction and Target Site Sequencing. The 16 canavanine-resistant *S. cerevisiae* BY4741 colonies (Figure S4) or 7 randomly picked *S. cerevisiae* CEN.PK2-1c colonies (Figure 5) were each inoculated into 2 mL of YPAD media and cultured overnight at 250 rpm, 30 °C. One milliliter of overnight culture was spun down, and genomic DNA was extracted using the Wizard genomic DNA purification kit (Promega, San Luis Obispo, CA). The target sites were PCR-amplified and sequenced by Sanger sequencing (ACGT, Inc., Wheeling, IL).

■ ASSOCIATED CONTENT

Supporting Information

Additional methods, figures, and tables as described in the text. This material is available free of charge via the Internet at http://pubs.acs.org.

■ AUTHOR INFORMATION

Corresponding Author

*Tel: +1 217 333 2631; Fax: +1 217 333 5052; E-mail: zhao5@illinois.edu.

Present Address

⊥(N.S.) Department of Chemistry and Chemical Biology, Harvard University, 12 Oxford Street, Cambridge, Massachusetts 02138, United States.

Author Contributions

‖Z.B. and H.X. contributed equally to this work. H.X., J.L., Z.B., and H.Z. designed the experiments; Z.B. and H.X. performed all of the experiments with the help of L.Z., X.X., N.S., and T.S.; H.X., Z.B., and H.Z. wrote the manuscript.

Notes

The authors declare no competing financial interest.

■ ACKNOWLEDGMENTS

This work was supported by the Center for Industrial Biotechnology at the University of Illinois at Urbana–Champaign. The authors thank Dr. Jennifer A. Doudna (Department of Molecular and Cell Biology, University of California, Berkeley, CA, USA) for sharing plasmid pMJ806, Dr. Feng Zhang (Department of Biological Engineering, Massachusetts Institute of Technology (MIT), Cambridge, MA, USA) for sharing plasmid pX260, and Dr. Ryan E. Cobb (Department of Chemical and Biomolecular Engineering, University of Illinois at Urbana–Champaign, Urbana, IL 61801, USA) for helpful suggestions.

■ ABBREVIATIONS

CRISPR–Cas, Clustered Regularly Interspaced Short Palindromic Repeats and CRISPR-associated proteins; DSB, double strand break; gRNA, guide RNA; HR, homologous recombination; PAM, protospacer adjacent motif; TALENs, transcription activator-like effector nucleases; tracrRNA, trans-encoded RNA

■ REFERENCES

(1) Yu, B. J., Sung, B. H., Koob, M. D., Lee, C. H., Lee, J. H., Lee, W. S., Kim, M. S., and Kim, S. C. (2002) Minimization of the *Escherichia coli* genome using a Tn5-targeted Cre/loxP excision system. *Nat. Biotechnol. 20*, 1018–1023.

(2) Komatsu, M., Uchiyama, T., Omura, S., Cane, D. E., and Ikeda, H. (2010) Genome-minimized *Streptomyces* host for the heterologous expression of secondary metabolism. *Proc. Natl. Acad. Sci. U.S.A. 107*, 2646–2651.

(3) Parsons, A. B., Brost, R. L., Ding, H., Li, Z., Zhang, C., Sheikh, B., Brown, G. W., Kane, P. M., Hughes, T. R., and Boone, C. (2004) Integration of chemical-genetic and genetic interaction data links bioactive compounds to cellular target pathways. *Nat. Biotechnol. 22*, 62–69.

(4) McCarroll, S. A., Bradner, J. E., Turpeinen, H., Volin, L., Martin, P. J., Chilewski, S. D., Antin, J. H., Lee, S. J., Ruutu, T., Storer, B., Warren, E. H., Zhang, B., Zhao, L. P., Ginsburg, D., Soiffer, R. J., Partanen, J., Hansen, J. A., Ritz, J., Palotie, A., and Altshuler, D. (2009) Donor-recipient mismatch for common gene deletion polymorphisms in graft-versus-host disease. *Nat. Genet. 41*, 1341–1344.

(5) Raab, A. M., Gebhardt, G., Bolotina, N., Weuster-Botz, D., and Lang, C. (2010) Metabolic engineering of *Saccharomyces cerevisiae* for the biotechnological production of succinic acid. *Metab. Eng. 12*, 518–525.

(6) Atsumi, S., Cann, A. F., Connor, M. R., Shen, C. R., Smith, K. M., Brynildsen, M. P., Chou, K. J., Hanai, T., and Liao, J. C. (2008) Metabolic engineering of *Escherichia coli* for 1-butanol production. *Metab. Eng. 10*, 305–311.

(7) Bennett, C. B., Lewis, L. K., Karthikeyan, G., Lobachev, K. S., Jin, Y. H., Sterling, J. F., Snipe, J. R., and Resnick, M. A. (2001) Genes required for ionizing radiation resistance in yeast. *Nat. Genet. 29*, 426–434.

(8) Khurana, V., and Lindquist, S. (2010) Modelling neurodegeneration in *Saccharomyces cerevisiae*: why cook with baker's yeast? *Nat. Rev. Neurosci. 11*, 436–449.

(9) Ostergaard, S., Olsson, L., and Nielsen, J. (2000) Metabolic engineering of *Saccharomyces cerevisiae*. *Microbiol. Mol. Biol. Rev. 64*, 34–50.

(10) Lorenz, M. C., Muir, R. S., Lim, E., McElver, J., Weber, S. C., and Heitman, J. (1995) Gene disruption with PCR products in *Saccharomyces cerevisiae*. *Gene 158*, 113–117.

(11) Akada, R., Kitagawa, T., Kaneko, S., Toyonaga, D., Ito, S., Kakihara, Y., Hoshida, H., Morimura, S., Kondo, A., and Kida, K. (2006) PCR-mediated seamless gene deletion and marker recycling in *Saccharomyces cerevisiae*. *Yeast 23*, 399–405.

(12) Davidson, J. F., and Schiestl, R. H. (2000) Mis-targeting of multiple gene disruption constructs containing hisG. *Curr. Genet. 38*, 188–190.

(13) Delneri, D., Tomlin, G. C., Wixon, J. L., Hutter, A., Sefton, M., Louis, E. J., and Oliver, S. G. (2000) Exploring redundancy in the yeast genome: an improved strategy for use of the cre–loxP system. *Gene 252*, 127–135.

(14) Suzuki, Y., St Onge, R. P., Mani, R., King, O. D., Heilbut, A., Labunskyy, V. M., Chen, W., Pham, L., Zhang, L. V., Tong, A. H., Nislow, C., Giaever, G., Gladyshev, V. N., Vidal, M., Schow, P., Lehar, J., and Roth, F. P. (2011) Knocking out multigene redundancies via cycles of sexual assortment and fluorescence selection. *Nat. Methods 8*, 159–164.

(15) Wang, H., Yang, H., Shivalila, C. S., Dawlaty, M. M., Cheng, A. W., Zhang, F., and Jaenisch, R. (2013) One-step generation of mice carrying mutations in multiple genes by CRISPR/Cas-mediated genome engineering. *Cell 153*, 910–918.

(16) Jiang, W. Y., Bikard, D., Cox, D., Zhang, F., and Marraffini, L. A. (2013) RNA-guided editing of bacterial genomes using CRISPR–Cas systems. *Nat. Biotechnol. 31*, 233–239.

(17) Hwang, W. Y., Fu, Y. F., Reyon, D., Maeder, M. L., Tsai, S. Q., Sander, J. D., Peterson, R. T., Yeh, J. R. J., and Joung, J. K. (2013) Efficient genome editing in zebrafish using a CRISPR–Cas system. *Nat. Biotechnol. 31*, 227–229.

(18) Cong, L., Ran, F. A., Cox, D., Lin, S. L., Barretto, R., Habib, N., Hsu, P. D., Wu, X. B., Jiang, W. Y., Marraffini, L. A., and Zhang, F. (2013) Multiplex genome engineering using CRISPR/Cas systems. *Science 339*, 819–823.

(19) Mali, P., Yang, L., Esvelt, K. M., Aach, J., Guell, M., DiCarlo, J. E., Norville, J. E., and Church, G. M. (2013) RNA-guided human genome engineering via Cas9. *Science 339*, 823–826.

(20) Dicarlo, J. E., Norville, J. E., Mali, P., Rios, X., Aach, J., and Church, G. M. (2013) Genome engineering in *Saccharomyces cerevisiae* using CRISPR–Cas systems. *Nucleic Acids Res. 41*, 4336–4343.

(21) Moerschell, R. P., Tsunasawa, S., and Sherman, F. (1988) Transformation of yeast with synthetic oligonucleotides. *Proc. Natl. Acad. Sci. U.S.A. 85*, 524–528.

(22) Yamamoto, T., Moerschell, R. P., Wakem, L. P., Ferguson, D., and Sherman, F. (1992) Parameters affecting the frequencies of transformation and co-transformation with synthetic oligonucleotides in yeast. *Yeast 8*, 935–948.

(23) Doyle, E. L., Hummel, A. W., Demorest, Z. L., Starker, C. G., Voytas, D. F., Bradley, P., and Bogdanove, A. J. (2013) TAL effector specificity for base 0 of the DNA target is altered in a complex, effector- and assay-dependent manner by substitutions for the tryptophan in cryptic repeat-1. *PLoS One 8*, e82120.

(24) Aouida, M., Piatek, M. J., Bangarusamy, D. K., and Mahfouz, M. M. (2014) Activities and specificities of homodimeric TALENs in *Saccharomyces cerevisiae*. *Curr. Genet. 60*, 61–74.

(25) Gasiunas, G., and Siksnys, V. (2013) RNA-dependent DNA endonuclease Cas9 of the CRISPR system: holy grail of genome editing? *Trends Microbiol. 21*, 562–567.

(26) Wang, T., Wei, J. J., Sabatini, D. M., and Lander, E. S. (2014) Genetic screens in human cells using the CRISPR–Cas9 system. *Science 343*, 80–84.

(27) Deltcheva, E., Chylinski, K., Sharma, C. M., Gonzales, K., Chao, Y., Pirzada, Z. A., Eckert, M. R., Vogel, J., and Charpentier, E. (2011) CRISPR RNA maturation by trans-encoded small RNA and host factor RNase III. *Nature 471*, 602–607.

(28) Karvelis, T., Gasiunas, G., Miksys, A., Barrangou, R., Horvath, P., and Siksnys, V. (2013) crRNA and tracrRNA guide Cas9-mediated DNA interference in *Streptococcus thermophilus*. *RNA Biol.* *10*, 841–851.
(29) Sun, N., Bao, Z. H., Xiong, X., and Zhao, H. M. (2014) SunnyTALEN: a second-generation TALEN system for human genome editing. *Biotechnol. Bioeng.* *111*, 683–691.
(30) Deveau, H., Garneau, J. E., and Moineau, S. (2010) CRISPR/Cas system and its role in phage-bacteria interactions. *Annu. Rev. Microbiol.* *64*, 475–493.
(31) Whelan, W. L., Gocke, E., and Manney, T. R. (1979) The CAN1 locus of *Saccharomyces cerevisiae*: fine-structure analysis and forward mutation rates. *Genetics* *91*, 35–51.
(32) Romanos, M. A., Scorer, C. A., and Clare, J. J. (1992) Foreign gene expression in yeast: a review. *Yeast* *8*, 423–488.
(33) Loison, G., Vidal, A., Findeli, A., Roitsch, C., Balloul, J. M., and Lemoine, Y. (1989) High level of expression of a protective antigen of schistosomes in *Saccharomyces cerevisiae*. *Yeast* *5*, 497–507.
(34) Dorfman, B. Z. (1969) The isolation of adenylosuccinate synthetase mutants in yeast by selection for constitutive behavior in pigmented strains. *Genetics* *61*, 377–389.
(35) Zwolshen, J. H., and Bhattacharjee, J. K. (1981) Genetic and biochemical properties of thialysine-resistant mutants of *Saccharomyces cerevisiae*. *J. Gen. Microbiol.* *122*, 281–287.
(36) Pagano, A., Castelnuovo, M., Tortelli, F., Ferrari, R., Dieci, G., and Cancedda, R. (2007) New small nuclear RNA gene-like transcriptional units as sources of regulatory transcripts. *PLoS Genet.* *3*, 174–184.
(37) Szczebara, F. M., Chandelier, C., Villeret, C., Masurel, A., Bourot, S., Duport, C., Blanchard, S., Groisillier, A., Testet, E., Costaglioli, P., Cauet, G., Degryse, E., Balbuena, D., Winter, J., Achstetter, T., Spagnoli, R., Pompon, D., and Dumas, B. (2003) Total biosynthesis of hydrocortisone from a simple carbon source in yeast. *Nat. Biotechnol.* *21*, 143–149.
(38) Nishimasu, H., Ran, F. A., Hsu, P. D., Konermann, S., Shehata, S. I., Dohmae, N., Ishitani, R., Zhang, F., and Nureki, O. (2014) Crystal structure of Cas9 in complex with guide RNA and target DNA. *Cell* *156*, 935–949.
(39) Young, L. S., Rivier, D. H., and Sprague, K. U. (1991) Sequences far downstream from the classical tRNA promoter elements bind RNA polymerase III transcription factors. *Mol. Cell. Biol.* *11*, 1382–1392.
(40) Agmon, N., Pur, S., Liefshitz, B., and Kupiec, M. (2009) Analysis of repair mechanism choice during homologous recombination. *Nucleic Acids Res.* *37*, 5081–5092.
(41) Chen, Y., Partow, S., Scalcinati, G., Siewers, V., and Nielsen, J. (2012) Enhancing the copy number of episomal plasmids in *Saccharomyces cerevisiae* for improved protein production. *FEMS Yeast Res.* *12*, 598–607.
(42) Jinek, M., Chylinski, K., Fonfara, I., Hauer, M., Doudna, J. A., and Charpentier, E. (2012) A programmable dual-RNA-guided DNA endonuclease in adaptive bacterial immunity. *Science* *337*, 816–821.
(43) Liang, J., Chao, R., Abil, Z., Bao, Z., and Zhao, H. (2014) FairyTALE: a high-throughput TAL effector synthesis platform. *ACS Synth. Biol.* *3*, 67–73.
(44) Gietz, R. D., and Schiestl, R. H. (2007) High-efficiency yeast transformation using the LiAc/SS carrier DNA/PEG method. *Nat. Protoc.* *2*, 31–34.

376792 Uncovering Structure/Function Design Principles of RNA Transcriptional Attenuators Using Shape-Seq to Characterize RNA Structures in Vivo

Abstract:

RNA gene regulators are emerging as powerful parts for engineering gene regulation and networks. A potential advantage to using RNA regulators over their protein counterparts is the ability to uncover structure/function relationships and infer functional design principles using next generation tools that have recently been developed to characterize RNA structures *in vivo* (1). A particularly versatile RNA regulator is the pT181 transcriptional attenuator - an RNA sequence in the 5' untranslated region of a transcript that regulates transcription elongation in response to an antisense RNA expressed in *trans*. In previous work, a library of orthogonal transcription attenuators was engineered by replacing the antisense-RNA sensing region of the pT181 transcriptional attenuator with the sensing regions from five different naturally occurring antisense-RNA translational regulators (2). The previous study produced a number of regulators that varied in attenuation capability and orthogonality. Some hypotheses were made about how the predicted secondary structures of the attenuators caused the variation, though we had no quantitative method to assess these structure/function relationships. In this work, we chemically probe the structures of these regulators using a newly developed in-cell SHAPE-Seq technique (1) in order to understand the structure/function relationship of these regulators. In-cell SHAPE-Seq provides nucleotide-resolution chemical reactivity spectra for RNAs that reflect their structural state within the cell. By probing the attenuator, antisense, and interacting attenuator-antisense RNA's of both functional and non-functional regulators, we show that we can directly uncover important structural features required for proper transcription attenuation. This knowledge builds our understanding of how these RNAs function, and brings us closer to a design paradigm that uses experimentally derived structure/function relationships to design chimeric attenuators *in silico*. Finally, our approach of combining functional experiments with in-cell SHAPE-Seq as a tool for understanding the connection between RNA structure and function is completely general, and represents a quantitative approach to rationally engineering regulatory RNAs.

(1) **Watters K. E., et al.**, "In-cell RNA Structure Probing with SHAPE-Seq," submitted to the 2014 AIChE National Meeting (TA006), Atlanta, GA (Nov 16-21, 2014).

(2) **Takahashi M. K., and J. B. Lucks**, "A Modular Strategy for Engineering Orthogonal Chimeric RNA Transcription Regulators," *Nucleic Acids Res*, **41** (15) pp. 7577-7588 (Aug. 2013).

363647 Development of Heat-Repressible RNA Thermosensors in Bacteria

Abstract:

Inducible gene expression systems are powerful tools for optimizing metabolic pathways or controlling gene regulatory networks in organisms. However, chemical inducers must be added exogenously, leading to extra costs and potential barriers to scale-up. Expression systems that are induced by environmental changes avoid the need for costly chemical additions and allow bacteria to respond directly to growth conditions. The overall goal of this research is to develop small, synthetic, heat-repressible RNA thermosensors. These thermosensors are located in the 5' UTR upstream of the Shine-Dalgarno site, where they contain a cleavage site for ribonuclease E, an enzyme native to *Escherichia coli* that binds at the cleavage site and degrades the mRNA. At low temperatures, the recognition site is sequestered in a stem-loop structure. At high temperatures, the stem-loop unfolds and exposes the recognition site, and thus the mRNA is degraded, turning off gene expression. Synthetic heat-repressible RNA thermosensors were designed and tested *in vivo* by varying length, number of recognition sites, size of the loop, and predicted melting temperature of the stem. Cells containing a thermosensor-controlled GFP as well as an internal control RFP were grown at a variety of temperatures ranging from 15'C to 37'C, with the best performing thermosensor showing greater than 700-fold change in expression from low to high temperatures. These temperature-repressible thermosensors are small, do not require expression of protein regulators, and can function independently of expensive chemical inducers. In addition, they are simple to design and can be implemented in the optimization of a range of cellular processes.

373292 Engineering and Understanding Enzyme and Pathway Localization to ER and Lipid Droplet Membranes

Abstract:

Many essential and industrially important metabolic pathways are assembled on cellular membranes: proteins are glycosylated through a series of membrane-bound reactions; in plants isoprenoids are synthesized in well-organized multienzyme structures bound to intracellular membranes; and, the key step in acetate ester production in yeast is localized to the membrane surface of lipid droplets. Such localization is advantageous. High local concentrations of enzymes and reaction intermediates can increase pathway reactions rates and prevent undesired side reactions from consuming pathway intermediates, thus leading to increased pathway yields. Intracellular scaffolding technologies (e.g., protein and nucleic acid scaffolds) have re-created these advantages in engineered pathways by co-localizing enzymes in the cytosol. While these technologies have produced substantial gains in pathway yields they cannot readily accommodate membrane-bound enzymes. This gap in synthetic biology tools significantly limits our abilities to access the benefits of enzyme co-localization with membrane-bound pathways.

The goal of this work is to identify protein parts that can be used to direct the assembly to biosynthetic pathways comprised of natively membrane-bound and cytoplasmic enzymes on intracellular membranes. Specifically, we aim to develop synthetic biology tools that can sort endoplasmic reticulum (ER) enzymes to lipid droplets (LDs) and that can direct cytoplasmic enzymes to LDs. In this way, we aim to create co-localized reaction centers in the intracellular space. The first step in engineering such a system is to understand and control localization to LDs. For example, alcohol acetyltransferase-1 in *S. cerevisiae*, which catalyzes the condensation of simple alcohols and acetyl-CoA to the corresponding ester, sorts from the ER to lipid droplets where it remains during stationary phase. Through a series of structure-function studies we have identified dual N- and C-terminal amphipathic helices that are responsible for this intracellular localization. Our studies of LD localization extend to a series of protein tags known to function in mammalian and plant cells. We demonstrate that the terminal hydrophobic domain from a putative methyltransferase AAM-B can localize cytoplasmic proteins to LDs via the ER. Under fermentation conditions, we demonstrate that the LD protein olesin can be used as a terminal fusion to localize proteins to LDs in yeast. Building from these localization studies we aim to organize LD-localized multistep reaction pathways for the synthesis of acetate and ethyl esters.

385672 Regulating Malonyl-CoA Metabolism Via Synthetic Antisense RNAs for Enhanced Biosynthesis of Natural Products

Abstract:

Malonyl-CoA is not only a critical precursor for fatty acid biosynthesis, but also serves as an essential building block of natural products such as coumarins, flavonoids, and other polyketides. Its biosynthesis is tightly regulated and its intracellular concentration is usually controlled at low levels, which poses a great challenge to biosynthesize malonyl-CoA derived molecules via metabolic engineering. Previous efforts to increase malonyl-CoA availability were mainly focused on the overexpression of acetyl-CoA carboxylase, the enzyme responsible for the conversion of acetyl-CoA into malonyl-CoA. However, inactivation of its consumption pathway has not been achieved since the disruption of fatty acid biosynthesis would be lethal to the host cells. To overcome this limitation, we report an antisense RNA strategy to down-regulate fatty acid biosynthesis and reduce undesired malonyl-CoA consumption to support heterologous biosynthesis. To increase the stability of the antisense RNAs, we employed an artificial loop-stem structure, which was further optimized in terms of its length and binding position to maximize the interference efficiency. We observed the interference effects at transcriptional, translational, and metabolic levels. Furthermore, we used three heterologous biosynthetic pathways of polyketide compounds to test the applicability of this RNA-mediated down-regulating strategy. The results showed that this strategy greatly improved the biosynthesis of malonyl-CoA derived natural products. This work demonstrates that antisense RNA is an efficient synthetic biology tool to increase microbial production of economically and pharmaceutically valued compounds.

358000 Challenges and Opportunities in Synthetic Biology

Abstract:

Synthetic biology is the deliberate design of novel or improved biological systems and organisms that draws on principles elucidated by biologists, chemists, physicists, and engineers (1). It is a rapidly growing area with broad applications in medical, chemical, food, and agricultural industries and particularly has become a powerful engine for the exponential growth of the white biotechnology industry. In this talk, I will briefly discuss the challenges and opportunities in synthetic biology and highlight our recent work on the development and application of novel synthetic biology tools for biotechnological applications. Specifically, I will discuss two projects including (a) discovery, characterization, and engineering of novel natural product biosynthetic pathways for drug discovery and development (2,3); (b) development of novel genome-scale engineering tools for strain development and cell line creation (4-5).

1. H. Zhao. Synthetic Biology: Tools and Applications. Elsevier, 2013. pp350.
2. Y. Luo, H. Huang, J. Liang, M. Wang, L. Lu, Z. Shao, R. Cobb, and H. Zhao. "Activation and Characterization of a Cryptic Polycyclic Tetramate Macrolactam Biosynthetic Gene Cluster." *Nature Communications*, 4: 2894 (2013).
3. Z. Shao, G. Rao, C. Li, Z. Abil, Y. Luo, and H. Zhao. "Refactoring the Silent Spectinabilin Biosynthetic Gene Cluster Using a Plug-and-Play Scaffold." *ACS Synthetic Biology*, 2, 662-669 (2013).
4. Z. Shao, H. Zhao, and H. Zhao. "DNA Assembler, an *in vivo* Genetic Method for Rapid Construction of Biochemical Pathways." *Nucleic Acids Research*, 37, e16 (2009).
5. T. Si, Y. Luo, Z. Bao, and H. Zhao. "RNAi-Assisted Genome Evolution in *Saccharomyces cerevisiae* for Complex Phenotype Engineering." *ACS Synthetic Biology*, in press.
6. J. Liang, R. Chao, Z. Bao, N. Sun, Z. Abil, and H. Zhao."FairyTALE: A High-throughput TAL Effector Synthesis Platform." *ACS Synthetic Biology*, 3, 67-73 (2014).

367582 Multi-Scale in Vivo Systems Analysis of Cell-Cell Communication and Signaling in Complex Disease

Abstract:

Complex diseases such as cancer, arthritis, diabetes, and endometriosis involve interactions of tissue cells with immune system cells via molecular communication processes, and resulting disease consequently arises from dysregulated signaling network activities governing pathological cell behaviors. We are employing combined experimental/computational systems biology approaches to understand these communication and signaling processes at multiple scales, integrating across multiple intracellular pathways, multiple extracellular factors, and multiple cell types, with emphasis on in vivo studies in animal models and human subjects. This talk will present an overview of our paradigmatic concept, technical approaches, and example applications.

In-cell RNA Structure Probing with SHAPE-Seq

Kyle E. Watters*, Timothy R. Abbott, Julius B. Lucks

The increased sophistication of synthetic RNA regulators has renewed interest in characterizing RNA structure/function relationships. This has ushered RNA structure measurement into the '-omics' era of biology with the advent of a host of emerging techniques that couple chemical structure probing with high-throughput, next-generation sequencing (NGS). **S**elective 2'-**H**ydroxyl acylation **A**nalyzed by **P**rimer **E**xtension **Seq**uencing (SHAPE-Seq) was the first such method to couple chemical modification of RNA with NGS to quantitatively measure the reactivity of each nucleotide in an RNA with rigorous statistics [1,2]. However, the original SHAPE-Seq work had only been used for *in vitro* probing experiments [1]. Recent advances by others [3,4] have shown that SHAPE reagents can be used to probe RNA structures as they exist in the cellular context, but were still hindered by the limitations of capillary electrophoresis [3] or gel electrophoresis [4].

In this work, we present in-cell SHAPE-Seq, a coupling between NGS and SHAPE probing in *E. coli*. We began the development of in-cell SHAPE-Seq by first creating a standardized platform for expressing regulatory RNAs that is capable of simultaneously measuring regulator function via a fluorescence assay and RNA structure with SHAPE-Seq. In-cell SHAPE-Seq performs chemical probing of RNAs directly in the cell, and then uses NGS to both: remove any off-target cDNAs generated, and probe multiple RNAs simultaneously. We show in-cell SHAPE-Seq's capabilities can be used to observe structural motifs in multiple interacting synthetic RNA regulators in *E. coli*, and can be used to directly observe RNA-RNA interactions in the cell. By providing quantitative RNA structural data for *in vivo* RNA folds, we anticipate in-cell SHAPE-Seq to motivate an entirely new approach to designing synthetic regulators that uses nucleotide-resolution structure/function relationships to uncover true RNA design principles. We expect that this type of characterization feedback at both the structure and function level will greatly speed the design of new regulators and become a staple of synthetic biology.

1. Lucks, J. B., et al. (2011) Multiplexed RNA structure characterization with selective 2'-hydroxyl acylation analyzed by primer extension sequencing (SHAPE-Seq). *PNAS*
2. Aviran, S., et al. (2011) Modeling and automation of sequencing-based characterization of RNA structure. *PNAS*
3. Tyrrell, J., et al. (2013) The cellular environment stabilizes adenine riboswitch RNA structure. *Biochemistry*
4. Spitale, R. C., et al. (2012) RNA SHAPE analysis in living cells. *Nature Chemical Biology*

376466 Systematic Analysis, Optimization and Extension of High-Throughput RNA Structure Probing with Shape-Seq

Abstract:

RNA structure is a primary determinant of its function, and there is great interest in methods that can characterize RNA structures in high-throughput. Recently, techniques that merge chemical probing of RNA structure with next generation sequencing have created breakthroughs in the throughput and scale of RNA structure characterization. However, little work has been done to examine the effects of library preparation and sequencing steps of these techniques on the measured structural information, which is critical for accurate data interpretation. Here, we present a thorough analysis and optimization of the SHAPE-Seq technique that combines Selective 2'-Hydroxyl Acylation Analyzed by Primer Extension (SHAPE), with next generation sequencing. We assess reactivity data generated by SHAPE-Seq over a benchmark panel of RNAs to that generated by the QuSHAPE capillary electrophoresis-based method. We also optimize the steps of SHAPE-Seq associated with library preparation and sequencing, and show that SHAPE-Seq provides highly reproducible nucleotide-resolution reactivity data over a wide range of structural contexts, with no apparent biases. We also present SHAPE-Seq v2.0, which uses a 'universal' reverse-transcriptase priming strategy to reveal reactivity information for every nucleotide in an RNA without having to use or introduce a specific reverse transcriptase priming site within the RNA sequence itself. We show that SHAPE-Seq v2.0 is also highly reproducible, and that incorporating SHAPE-Seq v2.0 reactivity data as constraints in structural prediction algorithms yields predictions on par with those generated using data from other SHAPE methods. This work demonstrates that the SHAPE-Seq methods do not bias the measurement of structure-dependent chemical reactivity data. We anticipate SHAPE-Seq v2.0 to be broadly applicable to understanding the RNA sequence-structure relationship at the heart of some of life's most fundamental processes.

392228 The Promises and Perils of Interpreting Ribosome Profiling Data

Abstract:

Ribosome profiling, a recently developed next-generation technique from the Weissman lab at UCSF, provides a "snapshot" of the numbers and locations of all mRNA-bound ribosomes throughout a cell's transcriptome. Ribosome profiling and complementary RNA-seq measurements have provided an unprecendented genome-wide view of the translation process, and how different mRNA sequence affects ribosome occupancies. As a consequence, ribosome profiling data has provided a new lens to study the effects of synonymous codon usage on translation elongation rates, to understand the role of post-transcriptional regulation in controlling protein expression, and to develop better design rules for engineering synthetic genetic systems.

Here, we describe a biophysical model of the overall translation process, including translation initiation, elongation, and termination, and show how its calculations are consistent with ribosome profiling data. From these calculations, we propose quantitative design rules for ensuring reliable protein expression, stabilizing mRNAs, and avoiding excess ribosome sequestration; these design rules depend non-linearly on both the mRNA's translation initiation rates as well as synonymous codon usages, and would be difficult to determine through trial-and-error random mutagenesis. We also highlight how ribosome occupancy data could be mis-interpreted to draw incorrect conclusions based on statistical correlations, particularly when ribosome occupancies are assumed to be proportional to translation rates. Thus, in conjunction with classic chemical engineering fundamentals, ribosome profiling offers several promising avenues to test our understanding of translation. The perils of relying only on statistical analysis are noted.

384151 Going Deep with Biomolecular Engineering and Design

Abstract:

This talk will highlight recent advances from my lab coupling DNA deep sequencing with biomolecular engineering and design. I will begin by describing a general method for the experimental determination of the single mutation fitness landscape of complete protein sequences. I will then show how this general method can be applied to multiple fields of biotechnological import. For antibody engineering, I will show how this method can be applied to finely map non-contiguous antibody epitopes that can be used to increase antibody specificity as well as predict escape mutants. For metabolic engineering, we have applied this method to map and optimize a catabolic pathway in living organisms for expedited processing of bio-oil. For diagnostic fields, I will show how this method can illuminate and switch the specificity determinants of binding proteins to different Gram-negative microorganisms, including *Yersinia pestis*. Perspective about further integrating fitness landscapes with computational design will also be discussed.

380677 Rnaseq Reveals Substrate-Specific Lignocellulosic Degradation Responses in Anaerobic Gut Fungi

Abstract:

Complex microbiomes within the digestive tract of herbivores are increasingly of interest for the discovery of novel lignocellulose degrading enzymes for the production of biofuels. Anaerobic fungi, in particular, are prolific digesters of crude lignocellulose through a combination of powerful secreted enzymes and invasive growth. However, due to difficulties associated with their isolation and study, very little is known about these enzymes. Here, we pursued a systems level strategy integrating transcriptomics and phenotypic studies to probe the hydrolytic capabilities of these unusual microbes in detail for the first time. Using a novel isolate of the *Piromyces* genus, we have assembled the first *de novo* strand specific gut fungal transcriptome with RNAseq revealing a complex landscape containing hundreds of novel biomass degrading enzymes and regulatory antisense transcripts. Many of these transcripts are strongly repressed by simple sugars and co-regulated within functional regulons of coordinated gene expression. Gene Set Enrichment Analysis also revealed a tunable hydrolytic response that was specific to each cellulosic substrate and core enzymes that are believed to be integral to substrate recognition and the induction of this response. Highly conserved fungal transcription factors from *Aspergillus* that may mediate this behavior were also identified. Our work captured a previously unelucidated layer of gut fungal regulation, provides a roadmap for future platform engineering efforts and illuminates successful natural strategies for the formulation of industrial cellulase cocktails.

387369 Integrated Transcriptome and Proteome Analysis in Clostridial Stress Response Model Unveils the Complexity behind the Genetic Program and Post-Transcriptional Gene Regulation

Abstract:

Clostridium acetobutylicum is the model organism for the acetone-butanol-ethanol (ABE) fermentation. These clostridial cells face a variety of stress due to the toxic metabolic products such acids (butyrate) and solvents (butanol) that are produced during fermentation. Several genome level transcriptional studies have shown the complexity behind the regulation of the stress response and tolerance in Clostridia that is driven by major changes in the genetic program. Yet, the post-transcriptional regulation can be comprehensively understood only using a quantitative proteomic approach. In this work, we performed an integrated analysis of the quantitative proteomic data obtained using isobaric tag for relative and absolute quantification (iTRAQ) and genome-wide transcriptional data obtained using microarrays and deep sequencing using Illumina HiSeq 2.0 at multiple time points and levels of stress to two major metabolites stressors – butanol and butyrate. Our results demonstrate the additional layer of post-transcriptional gene regulation combined with the role of small regulatory non-coding RNA (sRNA) in mediating these post transcriptional gene regulation. Particularly, the heat shock proteins that were found to be upregulated in the proteome data under stress in the WT strain, were found to be down regulated in *SolRH+pTAAD* strain (overexpressing the *sol* operon (*aad-ctfA-ctfB*) and the KO of the *solB* sRNA upstream of the *sol* operon) under same stress conditions. Longer 5' untranslated region (UTR) of these heat shock proteins revealed post-transcriptional gene regulation mediated by sRNAs and/or transcriptional factors.

This is the first comprehensive study that provides novel insights into the post-transcriptional gene regulation behind the complexity of stress response network, which can be targeted for practical purposes towards strain development for chemicals and fuels production.

Abstract

Candida glabrata (*C. glabrata*) is an important emerging opportunistic human fungal pathogen with an increasing incidence as the cause of both mucosal and systemic infections. In addition, the fungus is the most commonly used microorganism for pyruvate production and is considered to be a potential producer of many other fine chemicals and biofuels. The inherent high tolerance of *C. glabrata* to various environmental stressors play a key role in its life as a successful pathogen and an excellent producer. However, the underlying molecular mechanisms are not well-characterized and its ability to adapt to ever-changing environmental stressors from the perspective of evolution is not evaluated to the best of our knowledge. Here, we used the strategy of the adaptive laboratory evolution combined with next generation sequencing to study the genome plasticity of *C. glabrata* under different environmental stressors, hydrogen peroxide and heat, to uncover the important genetic determinants conferring the increased tolerance to the stressors. By sequencing the genome and analyzing the transcriptome of the time-course samples with enhanced tolerance isolated during the evolution we can shed some light on the adaptive landscapes under given environmental stressors. Taken together, these knowledge will benefit the engineering of *C. glabrata* to be a more successful producer of desired chemicals as well as the discovery of new drug targets for antifungal therapy against the infection caused by the pathogen.

366201 Assembly and Validation of Versatile Transcription Activator-like Effector Libraries

Abstract:

The ability to perturb individual genes in genome-wide experiments has been instrumental in unraveling cellular and disease properties. Here we introduce, describe the assembly, and demonstrate the use of comprehensive and versatile transcription activator-like effector (TALE) libraries. As a proof of principle, we built an 11-mer library that covers all possible combinations of the nucleotides that determine the TALE-DNA binding specificity. We demonstrate the versatility of the methodology by constructing a constraint library, customized to bind to a known p53 motif. To verify the functionality in assays, we applied the 11-mer library in yeast-one-hybrid screens to discover TALEs that activate human SCN9A and miR-34b respectively. Additionally, we performed a genome-wide screen using the complete 11-mer library to confirm known genes that confer cycloheximide resistance in yeast. Considering the highly modular nature of TALEs and the versatility and ease of constructing these libraries we envision broad implications for high-throughput genomic assays.

394222 Slow Growth Increases Myo-Inositol Availability for Glucaric Acid Production in S. Cerevisiae

Abstract:

Glucaric acid has been classified as a top value-added chemical from biomass for potential uses ranging from cancer therapies to sustainable nylons by the Department of Energy. A biosynthetic pathway was constructed in *E. coli* to convert glucose to glucaric acid but activity of the *Mus musculus* myo-inositol oxygenase (MIOX) enzyme was rate-limiting. New expression constructs enabled a one-step knock-in of the four pathway genes to *S. cerevisiae.* Protein sequence similarity mapping paired with simple structure-function hypotheses enabled efficient sampling of over 300 MIOX enzymes. Substituting a plant MIOX in the pathway increased glucaric acid titers detected in the YPD culture broth at 72hrs. Supplementation with 60mM myo-inositol tripled glucaric acid titers for yeast expressing either the plant MIOX (1.5 g/L) or the mouse MIOX (0.94g/L). However, myo-inositol supplementation did not affect glucaric acid titers in yeast expressing the mouse MIOX during slower growth on non-fermentable carbon sources such as glycerol or ethanol. Myo-inositol is a component of yeast membrane phospholipids and carbon-limited chemostat experiments demonstrated that slow growth rates lead to increased myo-inositol availability and glucaric acid titers. We pursued a fed-batch strategy in YPD with yeast expressing the plant MIOX and achieved final titers of 1.2g/L.

Development of Tissue Phantoms for a New Breast Cancer Detection Technique

Matthew A. Conrad
Caitlin B. Douglas
Adam J. Nolte, PhD

Rose-Hulman Institute of Technology | Department of Chemical Engineering

Background Information

- Healthy breast tissue has a Young's modulus of ~3 kPa and tumorous tissue has a Young's modulus of ~20 kPa

⟹ Gelatin hydrogels seem to be acceptably soft

- 250 Bloom B Gelatin was utilized

Rose-Hulman Institute of Technology | Department of Chemical Engineering

Research Goals

- To prepare gelatin tissue phantoms of a Young's modulus of healthy tissue with tumors of various sizes and geometries
- To characterize the mechanical properties such as the effects of aging, multiple tests, strain rate, and sodium alginate as an additive

Rose-Hulman Institute of Technology | Department of Chemical Engineering

Tissue Phantom Preparation

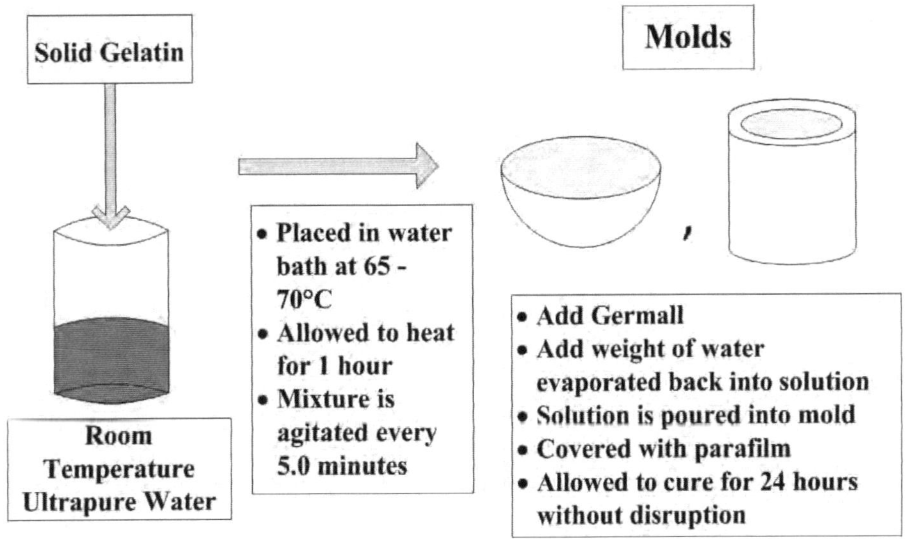

Rose-Hulman Institute of Technology | Department of Chemical Engineering

Tissue Phantom Preparation

Rose-Hulman Institute of Technology | Department of Chemical Engineering

Tumor Inclusion

Two proposed methods:

Liquid Injection via syringe

- 12.0% w/w gelatin liquid solution syringe-injected into congealing 4.0% w/w gelatin solution

⟹ • Not effective

Pre-made inclusions

- 12.0% w/w solid gelatin inclusion placed in congealing 4.0% w/w gelatin solution (T ~ 23.5°C)

⟹ • More effective approach

Rose-Hulman Institute of Technology | Department of Chemical Engineering

Mechanical Testing

Mechanical properties measured:

Young's Modulus
Fracture Strength
Toughness

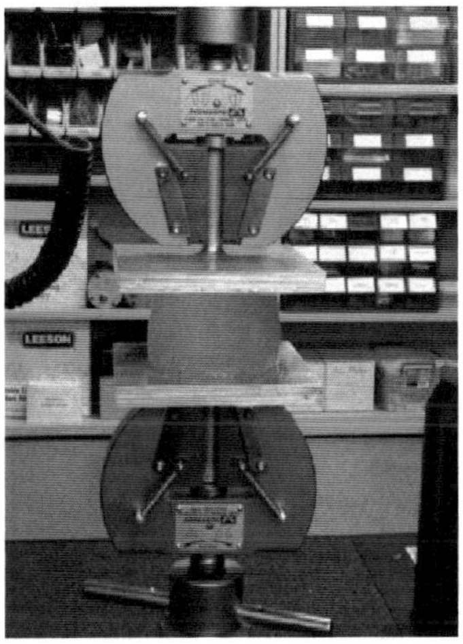

- Samples tested in compression with an Instron with a 10 kN LC
  - Young's Modulus from strains less than 10%
  - Some samples tested to failure

Rose-Hulman Institute of Technology | Department of Chemical Engineering

Concentration Dependence

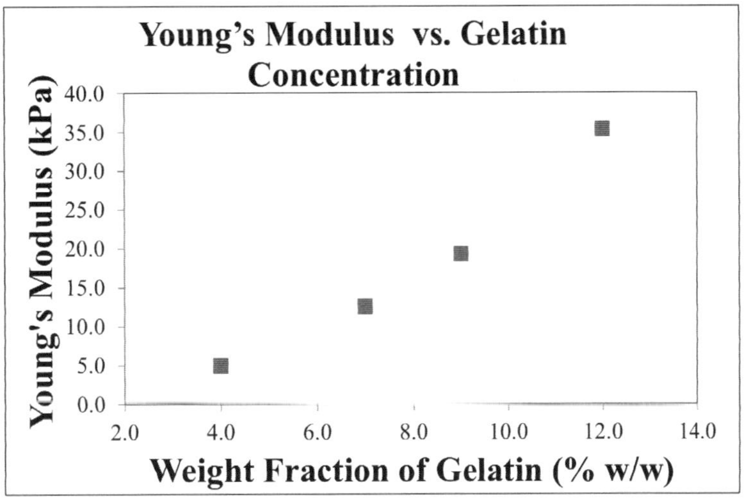

Rose-Hulman Institute of Technology | Department of Chemical Engineering

Aging Effects

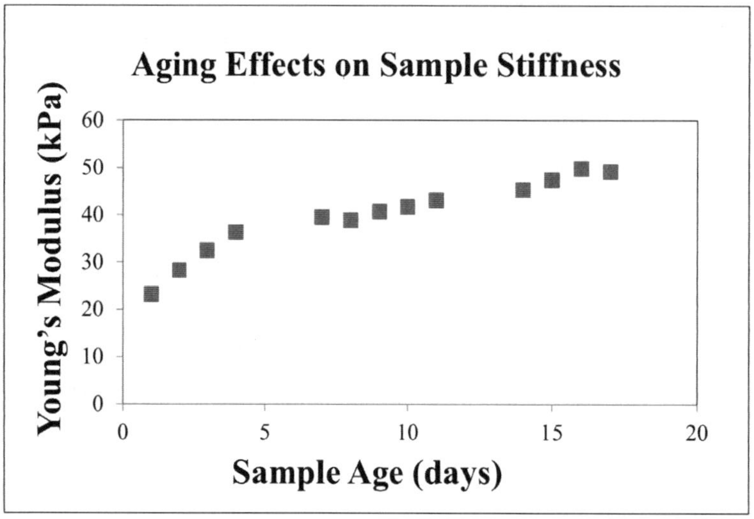

Rose-Hulman Institute of Technology | Department of Chemical Engineering

Testing Speed

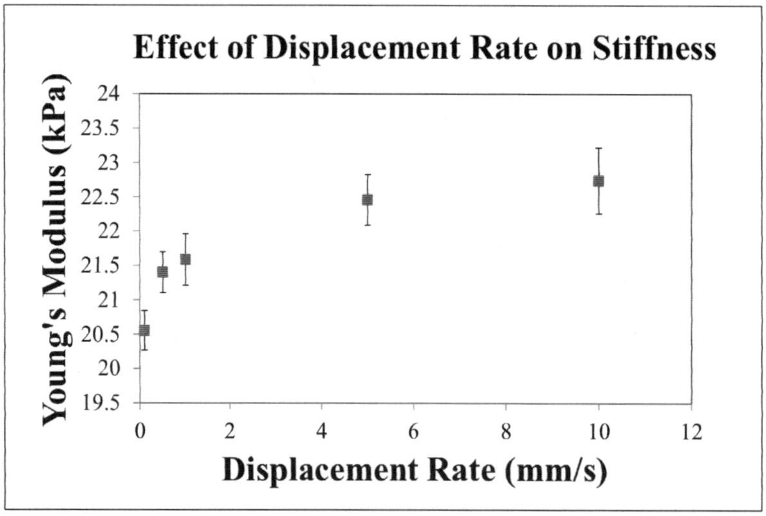

Rose-Hulman Institute of Technology | Department of Chemical Engineering

Multiple Test Effect on Stiffness

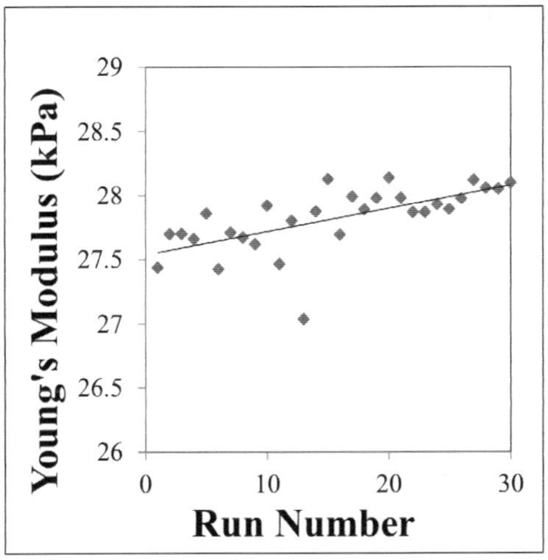

Uncovered during testing

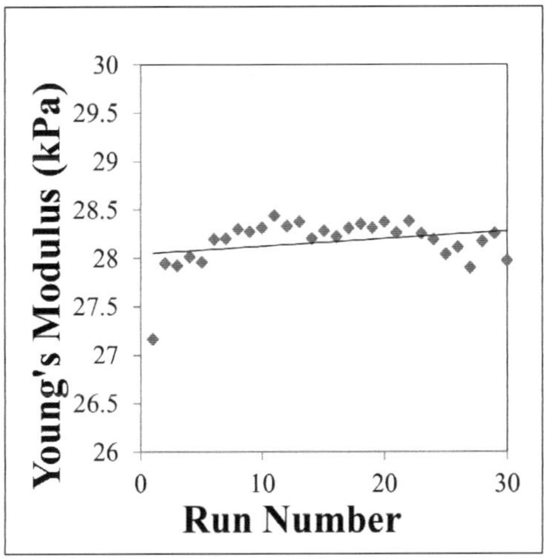

Covered during testing

Rose-Hulman Institute of Technology | Department of Chemical Engineering

Additive Effects: On Stiffness

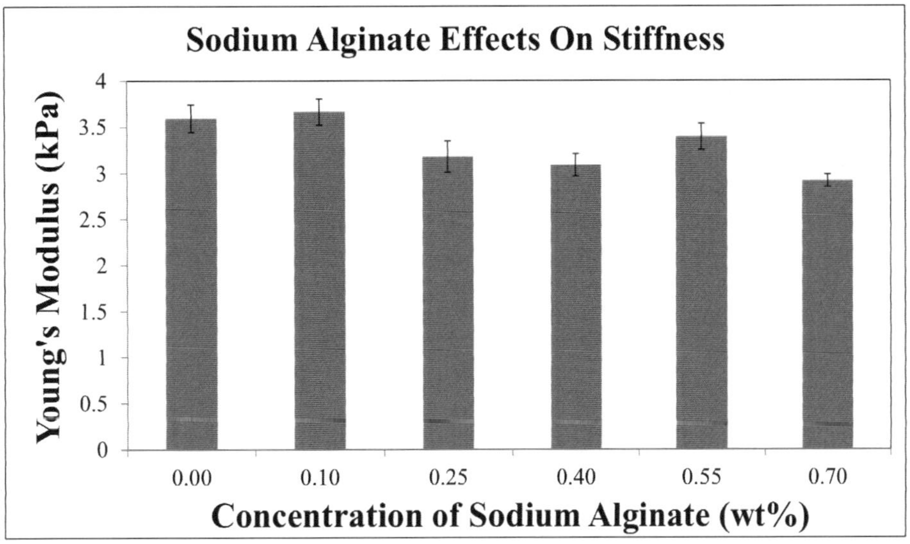

Rose-Hulman Institute of Technology | Department of Chemical Engineering

Additive Effects: On Fracture Strength

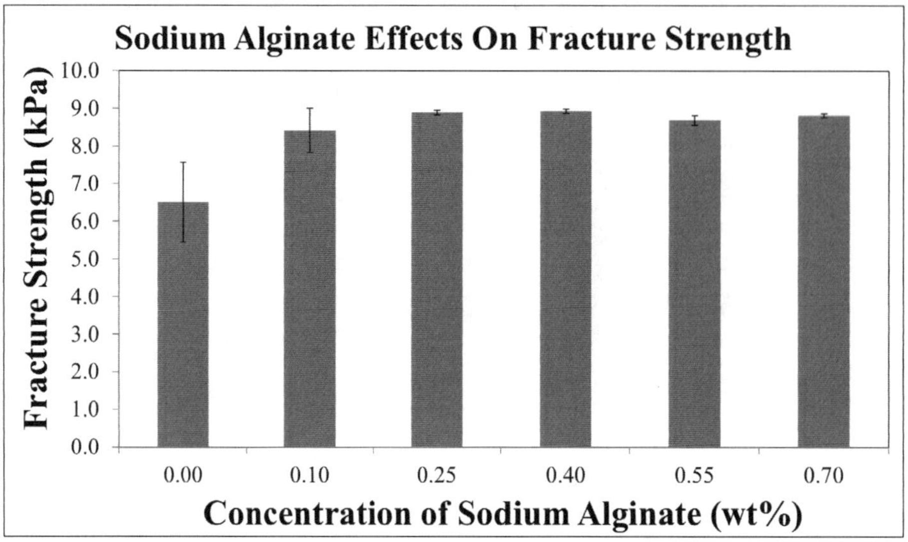

Rose-Hulman Institute of Technology | Department of Chemical Engineering

Additive Effects: On Toughness

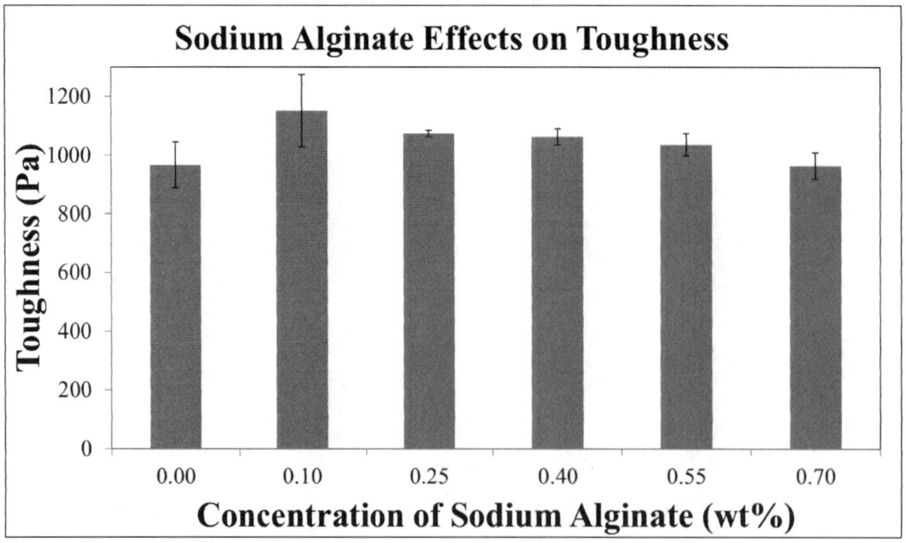

Rose-Hulman Institute of Technology | Department of Chemical Engineering

Conclusion

- Gelatin hemispheres can be made to mimic healthy breast tissue with tumor inclusions by inserting a pre-made tumor into a congealing gelatin solution
- Increasing the sodium alginate concentration appears to improve the fracture strength of gelatin hydrogels while decreasing the stiffness

Rose-Hulman Institute of Technology | Department of Chemical Engineering

Future

- Confirmation of results with texture analyzer
- More accurate relaxation tests
- Optimizing concentrations of additives

Rose-Hulman Institute of Technology | Department of Chemical Engineering

Acknowledgements

- *This research was funded by the National Science Foundation under Grant No. ECSS-1306808*

- *Professor Michael Insana and research group members (UIUC)*

- *Professors Olson and Throne (RHIT)*

Rose-Hulman Institute of Technology | Department of Chemical Engineering

Relaxation Effects

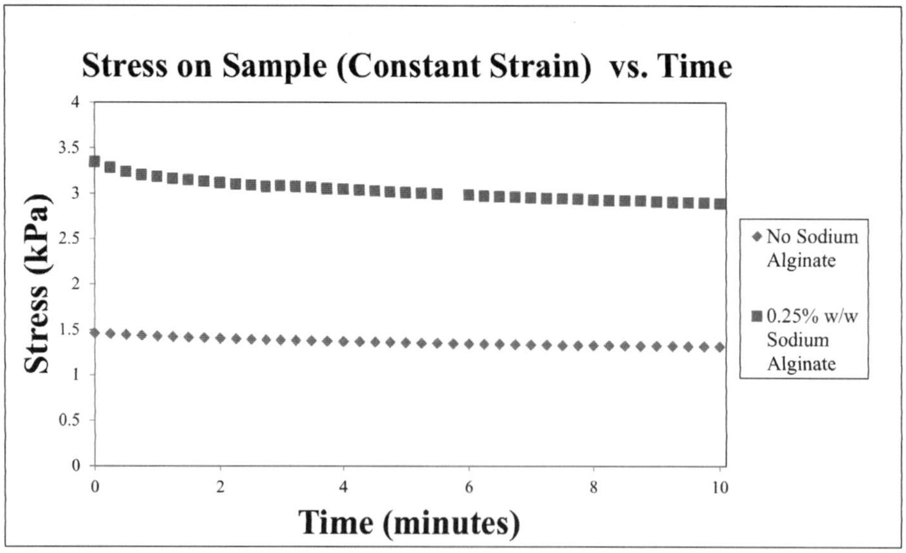

Rose-Hulman Institute of Technology | Department of Chemical Engineering

394192 Effective Antisense Design Using Ensemble of Energetically Sub-Optimal Secondary mRNA Structures

Abstract:

There is a deficit of effective therapies in the current research against bacterial infection. Many therapeutic strategies seek to use small molecules to target the infectious pathogen. One approach to disease inhibition involves direct manipulation of the pathogen at the RNA level. Part of the central dogma, messenger RNA (mRNA) is a transcript of genetic information that encodes the fundamental instruction for protein production. Inhibiting translation of mRNA effectively prevents synthesis of proteins.

In order to effectively block the RNA's message from being read, it must be physically accessed by the therapeutic agent. A field of study has arisen in which a complementary nucleic acid binding strand, called antisense[1], is generated to impede protein synthesis. One of the issues that arises in creating effective antisense RNA is the ability to find appropriate target sites on the mRNA to be inhibited. This problem is in part due to fluctuating secondary structure sequestering target sites. In fact, the kinetics of physical accession of the antisense to the target strand are suggested to be the rate-limiting factor[2] and thus the cause of inefficiency of this method. However, the fluctuations among suboptimal conformations are energetically inherent to the nucleic acid sequence and can be predicted. A software package, GenAVERT, has been developed to thermodynamically determine effective target sites by identifying the ensemble of the most probable energetically suboptimal states and determine which regions of mRNA are most accessible[3]. Once properly identified, a successful and effective antisense can, in theory, be synthesized.

In this work, a proof of principle will be demonstrated in *E. coli* with the downregulation of a short-lived variant of the green fluorescent protein through construction of an inducible antisense molecule. Statistical reduction in the fluorescence upon antisense induction will affirm the efficacy of this model.
Supported by fundamental energetic principles, this method of producing an antisense sequence complement for the most accessible mRNA target region has the potential to greatly reduce the strife of pathogenic infection.

[1] Lee, L. and Roth, C. (2003). Antisense technology in molecular and cellular bioengineering. *Current Opinion in Biotechnology*, [online] 14(5), pp.505-511.

[2] Patzel, V. and Sczakiel, G. (2000). In vitro selection supports the view of a kinetic control of antisense RNA-mediated inhibition of gene expression in mammalian cells. *Nucleic Acids Research*, [online] 28(13), pp.2462-2466.

[3] Johnson, E. and Srivastava, R. (2013). Volatility in mRNA secondary structure as a design principle for antisense. *Nucleic Acids Research*, [online] 41(3), pp.e43-e43.

Nanotechnology REU Summer 2014

Raghav Malik
AIChE National Conference: Undergraduate Research Forum II: Bioengineering, November 17th
Supervising Graduate Student: Giulia Suarato
Principal Investigator: Dr. Yizhi Meng
Department of Materials Science and Engineering

Stony Brook University

2

Proanthrocyanidin (PAC)

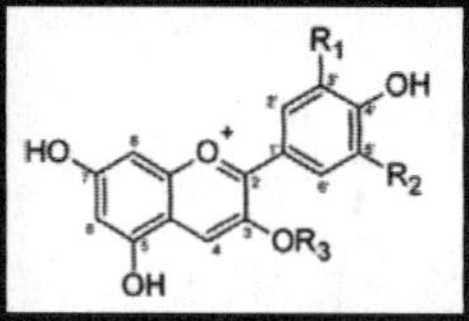

- Organically derived (from fruits and wines)
- Digests in saliva
- Promising chemotherapeutic effects, novel cancer cytotoxicity
- Encapsulation to improve bioavailability

Stony Brook University

Kom Kamonpatana, et. al. Susceptibility of anthocyanins to ex vivo degradation in human saliva, Food Chemistry. 2012: Vol. 135, 738-747

3

Chemotherapeutic Effects of PAC

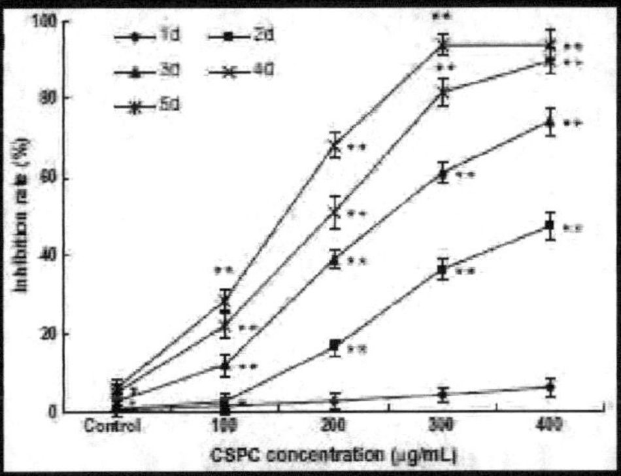

- Human Hepatoma G2 Cells
- Inhibition Rate: (Control – Treated / Control)*100
- Inhibition Rates Dose-Dependent
- Inhibition Time Dependent

NSF

Stony Brook University

H. Zhang, J. Ke, T. Shao, J. Li, Y. Duan, Y. He, C. Zhang, G. Chen, G. Sun, X. Sun. Food and Chemical Toxicology. 2014: Vol. 64, 166-176

4

Objective

Encapsulate PAC using chitosan and characterize the resulting nanoparticle as well as study the effect on MG-63 Osteosarcoma cells *in vitro*.

NSF

Stony Brook University

5

Why Chitosan?

- High Bioavailability
- Organic, derived from Chitin
- Biodegradable
- Hydrophilic

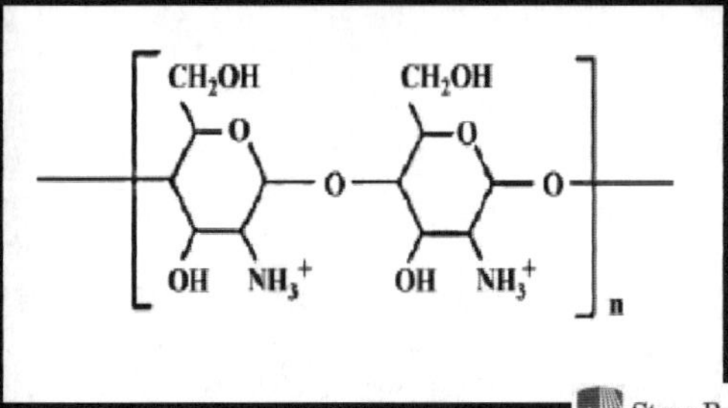

Chitosan

Stony Brook University

6

Encapsulation Potential

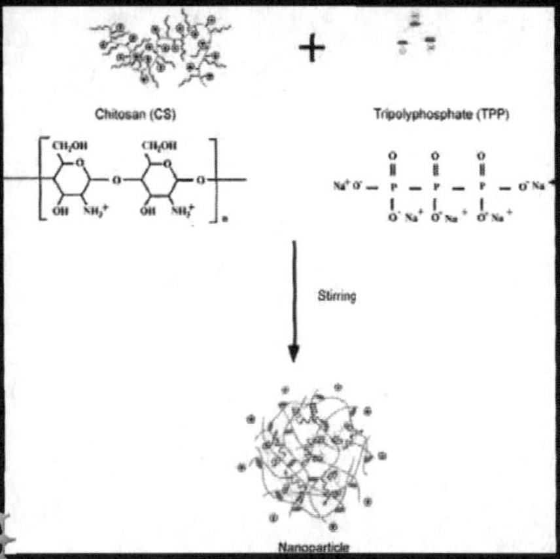

- Ionic Gelation Method
  - Pluronic F68 (detergent)
  - GCS:TPP 8.7:1 and 2.9:1
- Sizes are typically 200-300 nm

Stony Brook University

Adriana Trapani, Johannes Sitterberg, Udo Bakowsky, Thomas Kissel. International Journal of Pharmaceutics, 2009: Vol. 375, 97-106

7

Structure of Glycol Chitosan

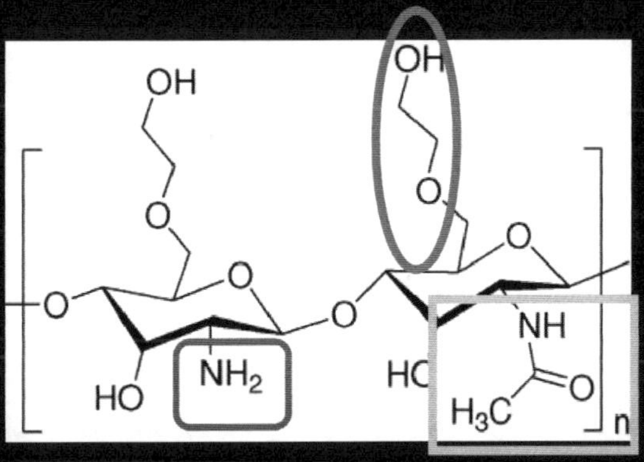

NSF

Stony Brook University

8

Approach

Synthesis

Characterization

Delivery *in vitro*

NSF

Stony Brook University

9

Fabrication and Characterization of NPs

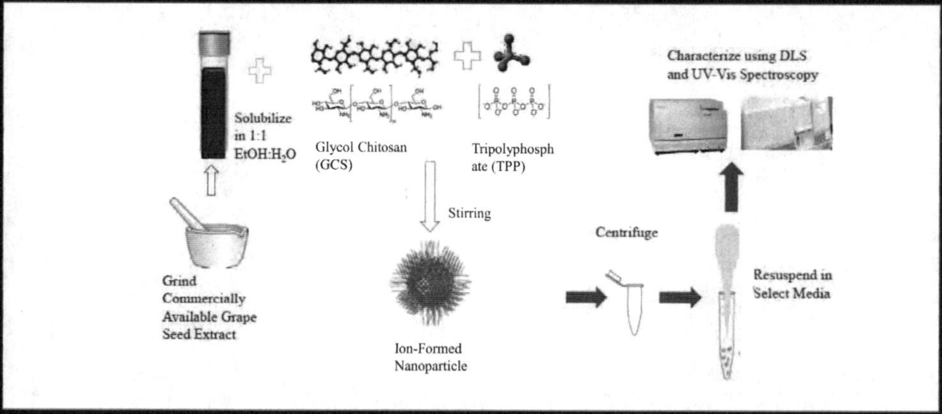

NSF

Stony Brook University

10

Nanoparticle Characterization

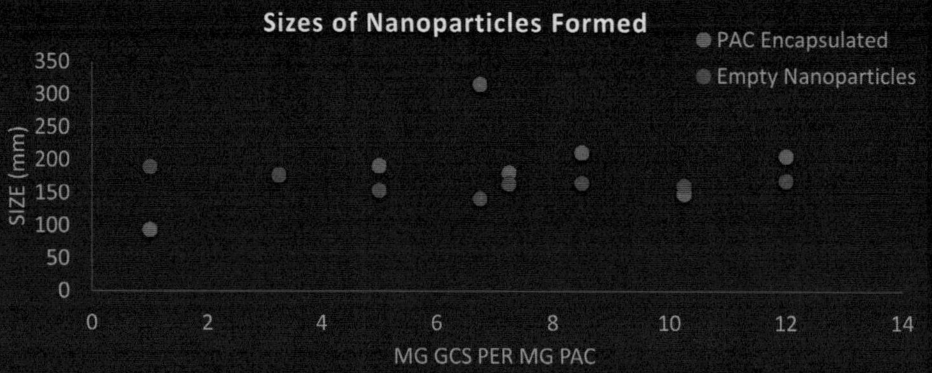

NSF

Stony Brook University

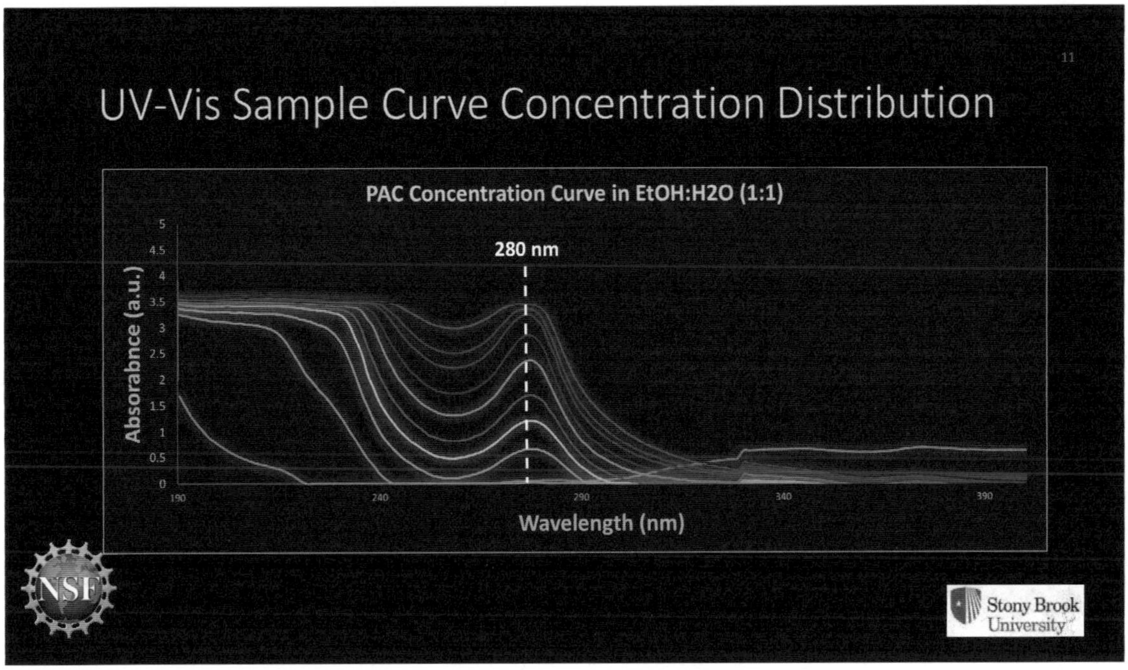
11
UV-Vis Sample Curve Concentration Distribution
PAC Concentration Curve in EtOH:H2O (1:1)
280 nm
Absorabnce (a.u.)
5
4.5
4
3.5
3
2.5
2
1.5
1
0.5
0
190
240
290
340
390
Wavelength (nm)
NSF
Stony Brook University

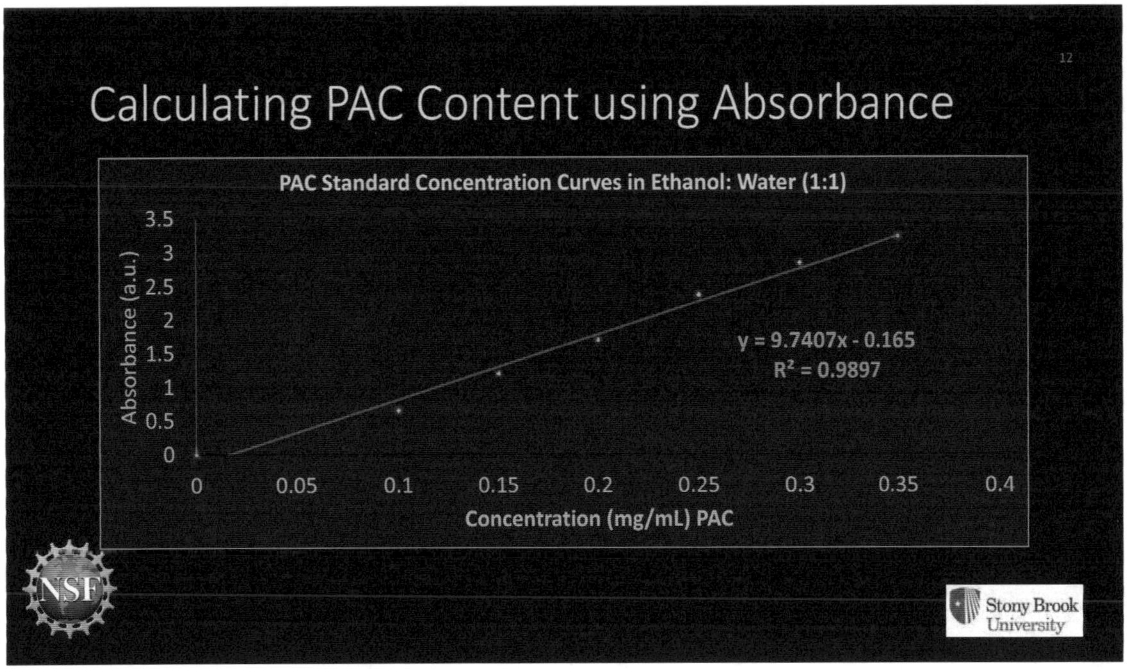
12
Calculating PAC Content using Absorbance
PAC Standard Concentration Curves in Ethanol: Water (1:1)
Absorbance (a.u.)
3.5
3
2.5
2
1.5
1
0.5
0
y = 9.7407x - 0.165
R² = 0.9897
0
0.05
0.1
0.15
0.2
0.25
0.3
0.35
0.4
Concentration (mg/mL) PAC
NSF
Stony Brook University

13

Encapsulation Efficiencies of GCS NPs

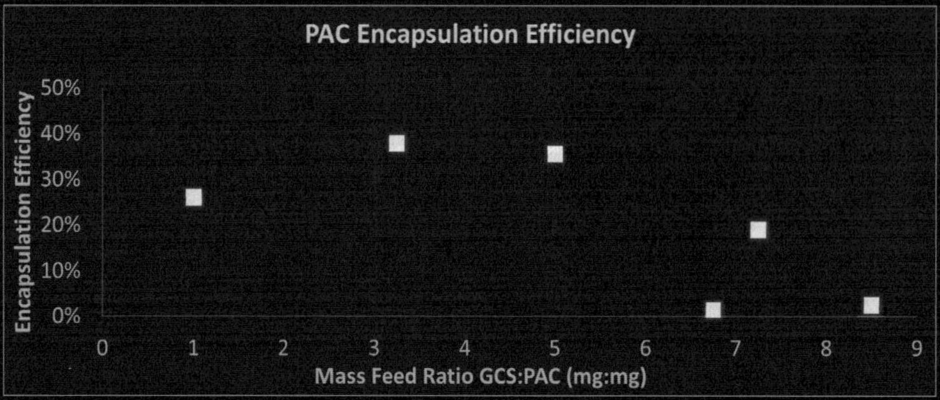

NSF

Stony Brook University

14

Delivery

- Cells used: MG-63 Human Osteosarcoma Cells
- Morphological study of cells after application
  - Comparing free and encapsulated PAC
- Cell Viability Assay done using MTS Assay
  - Comparing free and encapsulated PAC

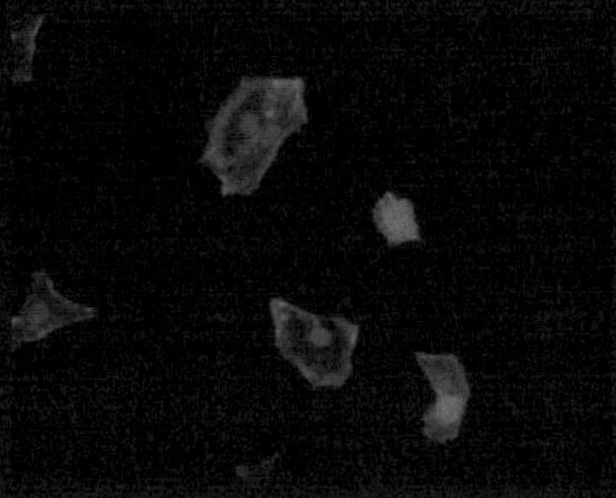

NSF

Stony Brook University

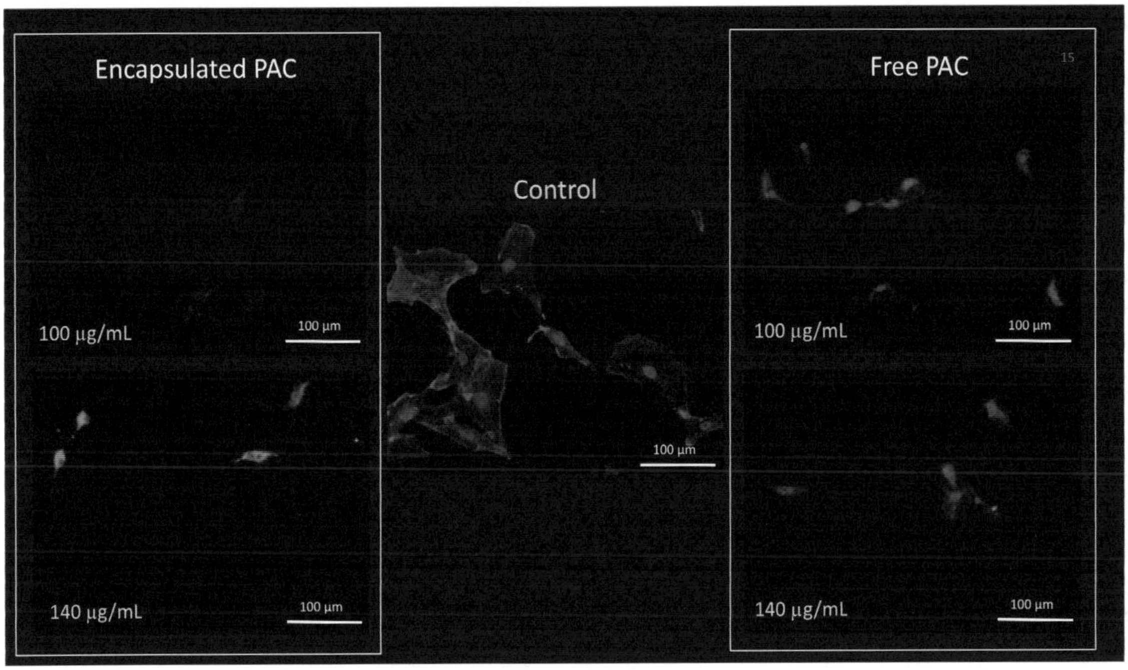

16

Cell Viability after 72 Hours

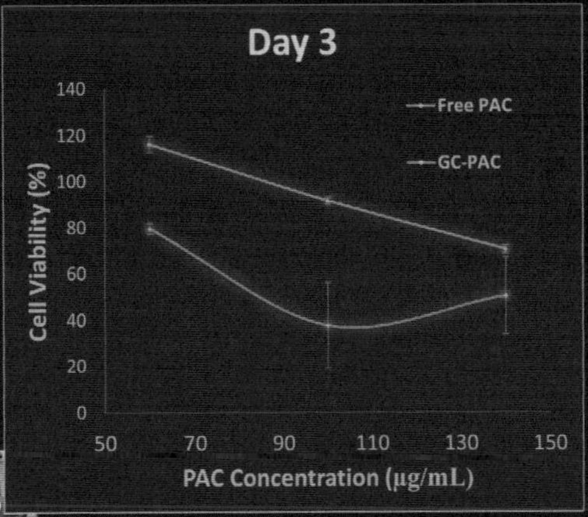

- MG-63 Osteosarcoma cells doubling time <72 hours
- Cell Viability is dose dependent
  - Can be adjusted for desired effect

Stony Brook University

17

Percent Cell Viability

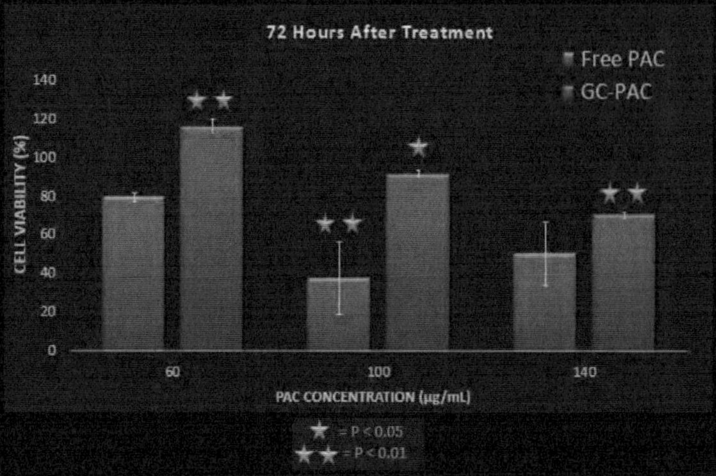

NSF

Stony Brook University

18

Conclusions

- Most Efficient Mass Feed Ratios: 5 mg : 1 mg GCS:PAC
- Morphological analysis shows similar effects by both free and encapsulated PAC
- MTS Analysis indicates encapsulated PAC is advantageous over free PAC in its dose-dependency as well as effectiveness with greater amounts of PAC

Future Studies

- Release Kinetics of GCS-PAC NPs
- Amount of GCS in formed NPs

NSF

Stony Brook University

19

Acknowledgements

I would like to thank the National Science Foundation, Giulia Suarato, Dr. Yizhi Meng, the Center for Inclusive Education, and all other students and mentors involved in the REU for making this research possible by providing the resources necessary

Stony Brook University

394191 Single Cell Isolation Via Microfluidic-Based Droplet Delivery

Abstract:

Commercial technologies such as fluorescent-activated cell sorting (FACS) for isolating and delivering single cell samples are often inappropriate for smaller research facilities due to their size and cost. Researcher labs conducting experiments on single cells require an inexpensive and accessible method for creating samples. We demonstrate the practicability of using a gravity-driven microfluidic device under a bench-top optical trap to generate droplets containing isolated single cells for delivery onto a 96 well plate. The microfluidic device dispenses droplets of constant volume at fixed intervals. We establish functionality by delivering droplets containing glutaraldehyde fixed red blood cells (RBCs) onto glass slides. A fluorescent confocal microscope capable of imaging the glutaraldehyde-induced fluorescence of the RBCs is employed to count the number of cells contained per droplet. We attained 64% fidelity of single-cell samples, indicating that delivery from our device is not stochastic. Biomedical researchers often expect droplet delivery methods to be compatible with a 96 well plate standard; to this end we achieved automated delivery into each well of a 96 well plate using a Teensy 3.0 encoded motorized stage. Additionally, we implemented a side and forward scatter detection scheme around the optical trap to show the feasibility of fully automating the cell isolation and droplet delivery process. The entire apparatus is compact and portable enough to fit within a sterile environment to prevent contamination of samples. We propose that our setup is an inexpensive and compact alternative to flow cytometers that is accessible to most labs and can reliably sort single cells into droplets for sterile delivery.

394190 Self-Assembling Zwitterionic Nanogels As Immune Isolating Coatings for Stem Cell Derived Pancreatic Islet Transplantation

Abstract:

The concept of pancreatic islet transplantation to patients with diabetes to restore normoglycemia is a viable clinical strategy. However, widespread clinical application of islet transplantation remains limited because of the harmful side effects of the necessary immunosuppressive therapy to prevent host rejection of transplanted cells. Here we report the fabrication of a superbiocompatible nanogel and a novel coating strategy for protection of stem cell derived pancreatic islets against host rejection in vivo. Zwitterionic nanogels (100-300nm in size) were fabricated using an inverse emulsion polymerization synthesis scheme and further surface modified with CLICK reactive functional groups. Stem cell derived pancreatic islet cells were engineered to express azide functionalities on their cell membranes. Confocal fluorescence microscopy was used to verify covalent reaction and assembly of dual-layered nanogel coatings on engineered islet cell clusters. Cytocompatibility of coating and islet cell functionality were examined using a combination of in vitro glucose tolerance tests, live/dead staining, and in vivo studies on mice treated with streptozocin, a small molecule drug that destroys the beta cells. Our results indicate that the developed nanogel formulation and encapsulation strategy provides a viable strategy for prolonged protection and isolation of transplanted islets against host rejection

394187 Modeling of Selenium Nanoparticle Deposition for Optimized Production of Antibacterial Surfaces

Abstract:

The Webster Nanomedicine lab designs and explores applications of nanomaterials to inhibit bacterial growth and cancer cell responses. Due to its beneficial effects on the body, and known chemopreventive and chemotherapeutic properties[1], selenium has been used in its nano-form to impart nano-roughness and chemical effects in these experiments. To ensure precise and consistent selenium nanoparticle (SeNP) coverage, we aim to tabulate several relevant aspects of the particle. By adjusting the parameters of the reaction producing SeNP, a variety of data can be collected and analyzed. Characteristics that are pertinent include the nanoparticle size, coverage, surface energy, and roughness. Eventually, a cubic response surface model will be developed using a central composite design (CCD), which will be able to predict the interaction of the particles with bacteria and cells.

SeNP are produced through a reaction of glutathione (GSH) with sodium selenite (Na_2SeO_3). The reactants are introduced in a 4:1 molar ratio, respectively. Thin discs of poly(l-lactic acid) (PLLA) are added to the reaction, providing a surface for the cells to grow. In order to precipitate out the SeNP, 2M NaOH is added to the mixture. The reaction is then halted by the addition of DiH_2O. The experimental design used four factors: time in GSH/ Na_2SeO_3, time in NaOH, reaction volume, and the number of samples in the volume. Three results are compared: nanoparticle size, nanoparticle coverage, and surface energy of each sample. Both the size and the coverage of the produced nanoparticles can be measured using scanning electron microscopy. The surface roughness is portrayed through atomic force microscopy. Surface energy can be calculated utilizing the contact angle, which is defined using a Goniometer.

Previous studies show longer time increments (in both cases) increase the deposition of SeNP on the substrate. However, as the increments reach larger values of time (30-60 seconds), there is less of a change in coverage. This suggests the existence of a saturation point. On the PLLA, there is a finite number of binding sites for the nanoparticles. The size of the nanoparticle, with time being the only deviation, remains consistent. It is predicted that development time and reaction volume directly correlate. Recent observations of previous studies have shown that increased coverage correlates with development time in both steps, and therefore, roughness.

Through characterization of the SeNP product, a cubic response surface will be calculated relating the input parameters to expected results, optimizing the coverage of nanoparticles on substrates. These efforts conserve both time and resources because reactions can be designed with focused specifications.

coverage. This suggests the existence of a saturation point. On the PLLA, there is a finite number of binding sites for the nanoparticles. The size of the nanoparticle, with time being the only deviation, remains consistent. It is predicted that development time and reaction volume directly correlate. Recent observations of previous studies have shown that increased coverage correlates with development time in both steps, and therefore, roughness.

Through characterization of the SeNP product, a cubic response surface will be calculated relating the input parameters to expected results, optimizing the coverage of nanoparticles on substrates. These efforts conserve both time and resources because reactions can be designed with focused specifications.

386306 Cell Cycle Transition in Embryonic Stem Cells during Self-Renewal and Differentiation

Abstract:

Embryonic stem cells are known to have a unique cell cycle, with a shortened G1 phase resulting in shorter doubling time. This G1 phase lengthens with differentiation, leading to an overall longer doubling time. A long-standing question in stem cell differentiation is the coupling of the cell cycle with differentiation, which still remains unanswered. Extensive experimental studies have offered interesting insights, in particular the establishment of the FUCCI reporter in human embryonic stem cells (hESCs). However, mechanistic information for this complex, non-linear interaction would be difficult to obtain with a purely experimental approach. Coupling experiments with mathematical modeling can aid in obtaining a more comprehensive understanding of the system. Our objective is to represent these unique dynamics of cell cycle transition in hESC using an integrated experimental and computational approach.

An important component in the analysis of this system is the heterogeneity in cell population and cell cycle state of each individual cell. Hence, we start off with a stochastic cellular automaton model to describe the behavior of the cells at the population level along with its associated heterogeneity. In this model, we track individual cells and their progress through the cell cycle, with the state of the cell being defined by the cell cycle phase which the cell is in (G1, S, or G2/M). The residence times for phases are described by probability distributions and their moments (e.g. normal distribution described by its mean and standard deviation). Individual cells are tracked in time, thereby simulating the dynamics of the cell population. Information from these dynamics can be obtained by comparing simulations to experimentally synchronized cells, achieved by arresting the cellular population, either hESC or hESC-derived pancreatic progenitor cells, in the G2/M phase with nocodazole.

Our population model, when applied to a self-renewing population of human pluripotent stem cells, is able to accurately predict phase resident time distributions from experimentally synchronized hESC. Interestingly, only one combination of specific distributions and their moments, representing phase residence times, were able to recapitulate the experimental data. These results were validated by a completely separate analysis, showing the accuracy of the modeling approach. In the next step this analysis was extended to explain the desynchronization dynamics of pancreatic progenitor phenotype. Fourier analysis predicts multiple subpopulations contributing to cell cycle desynchronization, and it was necessary to break the dynamics into these two subpopulations to capture the experimental behavior. The predicted phase distributions show that, in this more mature pancreatic progenitor phenotype, the G1 phases become both longer and more variable (as compared to undifferentiated cells).

To mechanistically explain this lengthening of G1 phase, we developed an ODE model of G1-S transition. We developed a skeletal model including key proteins reported to drastically change during differentiation: cyclin D, cyclin E, and p21 (CIP). Incorporation of these equation-based models in the automaton model results in a hybrid framework, where the G1 transition is governed purely by equations, while the S and G2/M phases are driven by the aforementioned probability distributions. The hybrid model is able to capture the overall trends of cell cycle transition and change in G1 population percentage during induced differentiation, in addition to the dependency on the differentiation molecule concentration. Furthermore, pertinent features of this transition, including dependency of lengthening on time and cell cycle position, probability of lengthening and maximum G1 residence time, were predicted.

Together, the developed single cell and stochastic population models can describe cell cycle behavior both at various stages of maturation as well as during differentiation. The population model and synchronization experiments are able to extract single cell information from simple population analyses, which can be mechanistically explained by the G1-S ODE model. By integrating the single cell model into the individual automatons of the population model, a more comprehensive understanding of the ESC cycle transition during differentiation is obtained.

372649 A New Regulator of Mitochondrial Metabolism That Has Different Functions in Spatially Distinct Subcellular Locations, the Mitochondria and the Nucleus

Abstract:

Mitochondrial energy production is closely regulated to conserve the cell's most valuable currency, ATP. Dysfunction in energy production and its regulation is associated with a variety of pathologic conditions that range from debilitating to major life-threatening diseases. Regulation includes communication between spacially distinct loci, in this case signaling from the nucleus to the mitochondria and retrograde regulation from the mitochondria back to the nucleus. Our understanding of stress-associated regulatory mechanisms for mitochondria in energy production and cell survival remains incomplete. We now report that we have discovered a new regulator of mitochondrial metabolism, the coiled-coil-helix-coiled-coil-helix domain-containing protein 2 (CHCHD2). It functions in a novel way by acting via different modes in the two cellular compartments. In normally growing cells most CHCHD2 is located in the mitochondria; during stress the amount of CHCHD2 in mitochondria decreases and the amount in the nucleus increases. In the mitochondria, CHCHD2 is imported by a Mia40-mediated pathway to the intermembrane space where, upon phosphorylation at Tyr-99, it binds to cytochrome *c* oxidase (COX), the terminal complex of the respiratory chain. This association is required for full COX activity; when CHCHD2 levels are lowered, such as by oxidative or hypoxic stress, widespread dysfunction is observed including reduction of COX activity, of membrane potential, and of growth rate whereas there are increased radical species (ROS) and mitochondrial fragmentation. In the nucleus, however, CHCHD2 acts as a transcription factor, one of whose targets is a COX subunit 4 isoform, *COX4i2,* which is transcriptionally stimulated by hypoxia and by the deacetylation of CHCHD2 by Sirt1. This CHCHD2-mediated stress response may provide an important survival mechanism for cells under conditions of oxidative or hypoxic stress, both in the acute phase by altering mitochondrial oxygen utilizing activity and in the chronic phase by promoting remodeling of COX. Supported by Henry L. Brasza endowment.

382411 Multi-Scale Control of Liver Regeneration: Integrating Molecular Regulation, Cell Phenotype Dynamics, and Physiological Response

Abstract:

Following damage, the liver initiates a recovery program inducing hepatocytes to enter the cell cycle and recover lost mass. As early as 30 seconds after injury, signaling cascades become active within the liver. Dynamic molecular changes continue for approximately 1 week following injury, restoring liver mass to the pre-injury levels. The dynamics of several molecular mediators of this process has been investigated previously to identify transient activation of inflammatory molecules for a few hours and a sustained activation of growth factors over several days. Yet much remains to be understood as to how the regulation of multiple molecular factors is coordinated to control liver repair mechanisms. Additionally, cell-types within the liver can each take on multiple distinct phenotypes either contributing to or inhibiting repair. The contributions of these phenotypes to liver repair and disease progression are just beginning to be appreciated. The present study takes a systems-based approach to investigate how multi-scale balances in cell phenotypes and molecular regulation impact liver regeneration.

We developed a computational model to synthesize the intrinsically multi-scale nature of liver regeneration by simulating connections between physiological-scale dynamics, activation phenotypes of non-parenchymal cells, and molecular signaling networks. Model analysis showed that shifting balances between populations of non-parenchymal cell activation phenotypes was sufficient to alter regeneration dynamics and overall tissue recovery following partial hepatectomy. As a perturbation to regeneration phenotype, we simulated alcohol-mediated suppression of liver regeneration by fitting our model to experimental data of liver recovery following chronic alcohol consumption and partial hepatectomy. Based on the model simulations, we predict that chronic alcohol consumption acts at a cellular-scale by shifting Kupffer cells from an M1 phenotype to an M2 phenotype following partial hepatectomy and by shifting hepatic stellate cells from a pro-regenerative phenotype to an anti-proliferative phenotype. At a molecular-scale, these changes in cell phenotypes are paralleled by dynamic increases in anti-inflammatory cytokine production and high levels of the anti-regenerative molecules such as fibrous collagens and TGFβ.

We tested these predictions at multiple-scales using high-throughput mRNA and protein measurements following partial hepatectomy in alcohol-fed rats and controls. We used laser capture microdissection to collect individual hepatic stellate cells from the livers of chronic alcohol-fed animals and controls before and after hepatectomy. We then used a high-throughput gene expression platform to quantify mRNA levels of ~100 genes across these individual cells and used multivariate statistics to identify clusters of hepatic stellate cell activation phenotypes. Our experimental results indicate that multiple activation of phenotypes of hepatic stellate cells arise following partial hepatectomy and alcohol exposure, consistent with model predictions. We followed a high-throughput ELISA-based approach to measure the cytokine microenvironment of the regenerating liver and how chronic alcohol consumption modulates the inflammatory milieu at the tissue-scale. Our results thus far indicate that anti-inflammatory cytokines increase in alcohol-fed animals prior to hepatocyte replication thus impairing hepatocyte priming, matching model predictions.

Taken together, these results provide novel insights into how molecular regulation during liver regeneration influences multiple scales to regulate non-parenchymal cell activation phenotype and tissue recovery.

Funding: R01 AA018873, T32 AA007463.

387189 A Biologically-Motivated Model of Chemotherapy-Induced Thrombocytopenia

Abstract:

A Biologically-Motivated Model of Chemotherapy-Induced Thrombocytopenia

Christine M. Carcillo[1], Robert S Parker[1,2,3,4]

1. Department of Chemical and Petroleum Engineering, Swanson School of Engineering 2. University of Pittsburgh Cancer Institute University of Pittsburgh 3. McGowan Institute of Regenerative Medicine 4. Clinical Research, Investigation, and Systems Modeling of Acute Illness (CRISMA) Laboratories, Department of Critical Care Medicine

Chemotherapy-induced thrombocytopenia is a hematological toxicity characterized by a thrombocyte, or platelet, count below average (<150,000 platelets/microliter). [1] Commonly employed drugs used against a variety of cancers (breast, small- and non-small-cell lung, ovarian, lymphoma, leukemia, and sarcoma) that lead to thrombocytopenia include doxorubicin, cyclophosphamide, ifosfamide, paclitaxel, carboplatin, cisplatin and etoposide. These chemotherapeutics kill stem and progenitor cells in the bone marrow, thereby decreasing the number of platelets released into circulation. With platelets having a lifespan of 7-10 days[1], the dynamic response to thrombocytopenic chemotherapy limits treatment. If toxicity is too severe, dose reductions must occur, thereby decreasing efficacy. By characterizing this toxic response to chemotherapy using a mathematical model, we can contribute to the design of patient-tailored treatment schedules for thrombocytopenic chemotherapeutics.

Mapping the biology onto mathematics, thrombopoiesis consists of a maturation chain that represents differentiation from a stem cell along its progression to progenitor cell, promegakaryocyte, and finally megakaryocyte, which bud off platelets into the blood stream. Once in circulation, about one-third of platelets sequester in the spleen, while the kidney, liver, and other organs and tissues also serve to sequester platelets at smaller fractions. Platelets are degraded by macrophages. Thrombopoietin (TPO), produced mainly in the liver and kidneys, is the primary regulator of thrombopoiesis, and it is inversely related to platelet production due to uptake of TPO by platelet and megakaryocyte TPO receptors. When platelet counts are low, the excess TPO stimulates the maturation chain. However, in TPO-deficient mice growth factors SDF-1 and FGF-4 were sufficient to increase platelet production post thrombocytopenia.[2] Hence, a more complex interactive network of positive and negative feedback is needed to maintain homeostasis. Once chemotherapy perturbs this homeostasis, a lag period of decreased platelet production will occur. An understanding of the life cycle of the platelet and the major regulatory players will lend insight into the recovery time after a specific chemotherapy challenge, as well as the potential interaction between different chemotherapy agents. At the systems scale, we use ordinary differential equations to describe the chemotherapeutic agent pharmacokinetic (PK) distribution and elimination. This, in turn, drives the biologically-motivated pharmacodynamic (PD) model of thrombocytopenia. As a result, the model is drug-specific, using the sigmoid Emax model parameters (Emax, EC50) to characterize the thrombocytopenic effect of a given agent, as well as interactions between agents.

Existing models of chemotherapy-induced thrombocytopenia have a phenomenological point of view, neglecting drug pharmacokinetics (PK) [Scholz et al. (2010)] or incorporating biological response and regulation via a nonlinear mathematical feedback effect [Bender et al. (2012)].[3,4] Our aim is to provide a more biologically relevant toxicity model that explicitly incorporates the effect of particular drugs – and their interactions – on the thrombocytopenia response.

As a case study, we examine the FDA-approved agent carboplatin. The PK model of Joerger et al. (2007) is used to drive the thrombocytopenia model, starting from a high dose of 800 mg/m^2 over 30 minutes.[5] Fitting parameters, including Emax and EC50 for carboplatin on thrombopoiesis, were adjusted to match model-derived platelet counts after carboplatin exposure to patient data from literature.[5, 6] Looking forward, combination therapy of paclitaxel and carboplatin has been shown to synergistically reduce thrombocytopenia when paclitaxel is administered first.[7] Modeling specific drugs and drug combinations can provide insight into dose scheduling, magnitude and duration. Additionally, cytokine treatments such as interleukin-1α can be administered to mitigate thrombocytopenia.[6] When employed in combination with a model-based treatment design algorithm, the thrombocytopenia model may contribute to patient-tailored dosing regimens that can reduce the number of severe toxicities encountered by cancer patients.

1. Sekhon SS, Roy V: Thrombocytopenia in adults: a practical approach to evaluation and management. S Med J 99:491-497, 2006

2. Avecilla, Scott T., et al. "Chemokine-mediated interaction of hematopoietic progenitors with the bone marrow vascular niche is required for thrombopoiesis." *Nature medicine* 10.1 (2004): 64-71.

3. Scholz, Markus, Arnd Gross, and Markus Loeffler. "A biomathematical model of human thrombopoiesis under chemotherapy." *Journal of theoretical biology* 264.2 (2010): 287-300.

4. Bender, Brendan C., et al. "A population pharmacokinetic/pharmacodynamic model of thrombocytopenia characterizing the effect of trastuzumab emtansine (T-DM1) on platelet counts in patients with HER2-positive metastatic breast cancer." *Cancer chemotherapy and pharmacology* 70.4 (2012): 591-601.

5. Joerger, Markus, et al. "Population pharmacokinetics and pharmacodynamics of paclitaxel and carboplatin in ovarian cancer patients: a study by the European organization for research and treatment of cancer-pharmacology and molecular mechanisms group and new drug development group." *Clinical Cancer Research* 13.21 (2007): 6410-6418.

6. Smith, John W., et al. "The effects of treatment with interleukin-1α on platelet recovery after high-dose carboplatin." *New England Journal of Medicine* 328.11 (1993): 756-761.

7. Xiong, Xiaoxiong, et al. "Cell cycle dependent antagonistic interactions between paclitaxel and carboplatin in combination therapy." *CANCER BIOLOGY AND THERAPY* 6.7 (2007): 1067.

383112 Quantification and Modeling of Infection Spread in the Presence of Virus-like Defective Interfering Particles

Abstract:

Defective interfering particles (DIPs) are virus mutants lacking genes essential for the production of viral proteins, but able to divert the viral proteins produced by an infectious virus, and thus, interfere with infectious virus production. Their ability to interfere with normal viral infections makes DIPs a useful tool to probe virus-cell interactions. Moreover, the accumulating evidence about their presence in natural infections and the increasing interest in vaccine formulations using DIPs highlight the need for understanding of the effect of DIPs on infection kinetics and on spread behavior of infection. In view of this, in this study, we analyzed the infection propagation behavior of a recombinant vesicular stomatitis virus (VSV) expressing red fluorescent protein (RFP) on cell monolayers in the presence of DIPs associated with the same virus strain. Quantification of infection spread behavior using multichannel fluorescence microscopy imaging and custom developed image processing workflow revealed the emergence of spread phenotypes with varying concentrations of DIPs. To further analyze the contribution of intracellular and extracellular processes involved in emergence of these different phenotypes, we developed a data-driven multilevel cellular automata model. The model incorporates the experimentally determined correlations between DIP input and single-cell level infection kinetics as well as the extracellular processes such as diffusion and adsorption, and therefore captures the stochasticity of the infection spread at multiple levels. Overall, this study provides insights about the interactions between viruses and DIPs that can guide future studies of the virally infectious disease progression in the presence of naturally occurring DIPs or DIPs administered as a vaccine component.

385290 Quantitative Analysis of Human Rhinovirus 2A Protease Disruption of Nucleocytoplasmic Trafficking Pathways and Effects on Infection Spread

Abstract:

Infections by human rhinovirus (HRV) contribute to a diversity of human illnesses that include the common cold, bronchitis, exacerbation of asthma and other chronic lung diseases. Previous studies have shown that HRVs grouped under three different species (HRV-A, B and C) span over 150 different genotypes and vary in their virulence and the severity of their infections. In parallel with clinical studies, genome sequencing has revealed a high level of diversity in protein sequences of the HRV 2A protease, an enzyme involved in viral polyprotein processing and shut off of host cell functions that can inhibit viral replication. Different 2A proteases target and cleave different nuclear pore proteins (Nups), and thereby disrupt nucleocytoplasmic signaling and transport pathways. This suggests that the proteases might have different enzymatic activities and carry-out their anti-host functions with unique mechanistic signatures. Here we compared the ability of different HRV 2A proteases to disrupt nuclear transport pathways using a novel cellular time-lapse fluorescent transfection assay. We created stable HeLa cells expressing fluorescent proteins fused to four different nuclear localizations signals, and we transfected these host cells with RNAs encoding different HRV 2A proteases. We observed the nuclear reporter protein accumulation in the cytoplasm/nucleus in real-time, and we employed a custom-developed java-based graphical interface to extract and analyze the fluorescent intensities from individually identified cells. This enabled us to quantify differences between HRV 2A proteases in their ability to disrupt nuclear transport pathways. We found that all proteases disrupted active import of a fluorescent reporter protein by the importin-α/β, transportin-1, and transportin-3 pathways as well as active export by the Crm1 pathway; however, different proteases inhibited these nuclear transport pathways with different efficiencies. We extended these studies to probe the spatial and temporal dynamics of spreading HRV infections using a dual color reporter system. Reporter HeLa cell monolayers were infected by fluorescent HRV strains where individual cells were tracked through the course of the infection; analysis of the spatial and temporal expression behaviors of individual cells was obtained for HRV strains expressing different 2A proteases, and these single-cell profiles were correlated with different spatial patterns of HRV infection spread. Overall, this work provides a foundation for linking how HRV diversity at the level of encoded 2A proteases may contribute to differential severity of HRV infections as they spread, which can guide the development of improved treatment strategies for HR-related illnesses.

369861 Computational Design of Hepatitis C Vaccines

Abstract:

Hepatitis C virus (HCV) is a global scourge afflicting 170 million people, nearly 2.5% of the human population. Chronic HCV infection is responsible for a quarter of all cases of liver cirrhosis and cancer, and is the leading cause of liver transplants in the developed world. The gold standard treatment of daily injections and oral tablets costs more than $15,000 and has failure rates of 30-50%, making it only moderately effective and essentially unavailable in the developing world. Prophylactic vaccination offers the most realistic and cost effective hope of controlling the epidemic.

Despite 20 years of research, a vaccine remains elusive. Progress has been frustrated in large part because we do not know which components of the immune system to activate with a vaccine. By combining models from statistical physics with machine learning and data mining of viral sequence databases, we devised an approach to infer quantitative models for viral fitness that reveal viral "soft spots" that are vulnerable to vaccine-induced immune pressure.

We have applied our approach to the RNA-dependent RNA polymerase of HCV genotype 1a that is responsible for replication of the viral genome. The predictions of our model are in good accord with experimental measurements and clinical data, including: 1) *in vitro* replicative fitness measurements, 2) clinically documented escape mutations, 3) intra-host viral evolution revealed by longitudinal deep sequencing, and 4) protective HLA-associated cytotoxic T-cell responses.

We use our model to computationally screen candidate vaccine immonogens to perform *in silico* design of promising cytotoxic T-cell immunogens that exploit the identified viral vulnerabilities and have broad coverage within the target population. By reducing the time and expense of trial-and-error testing, our computational design platform can inform and accelerate the search for a vaccine to arrest the HCV epidemic.

387738 Visualizing the Trajectory of Behavioral Changes in C. Elegans

Abstract:

Characterizing the progression of disease in an organism often requires integrating information from transient behaviors to slowly-changing attributes over a lifetime. Using the model organism *C. elegans*, we integrate high-dimensional phenotypic information on growth, posture, trajectory, and activity level across time-scales ranging from seconds to several days. Behavior on short time-scales is characterized using a hidden-Markov model. These states are used to create stereotyped trajectories of how populations of worms travel through behavior space during the first third of their adult lives. We show that perturbing the health of worms using environmental and genetic insults leaves a distinct impact on behavior worms engage in that persists over several days.

Design and Implementation of Whole-Cell Biosensors for Micronutrient Detection in Low Resource Environments

Micronutrient deficiencies are a significant healthcare concern across the globe. These conditions, though easily treated, remain a problem because they are often difficult to recognize and diagnose, requiring lab tests that are prohibitively expensive in both material and human resources for those in developing or remote areas. As obligate consumers of the same micronutrients, bacteria possess cellular machinery to control intracellular micronutrient levels and have corresponding regulatory mechanisms to respond to varying concentrations in their environment. We are developing a novel medical test based on bacterial sensors using designed genetic circuitry to direct existing or minimally engineered cellular machinery to trigger specific changes in color in response to defined micronutrient levels. Such a test would be cheap, require no complex equipment and minimal medical training to administer and interpret, allow on-site diagnosis of micronutrient deficiencies in the populations most at risk, and facilitate epidemiological studies of such deficiencies and assessment of the efficacy of intervention strategies on a population scale.

Towards this goal, we present work on a whole-cell biosensor for the detection of a micronutrient of medical relevance. We have demonstrated a multi-output circuit using two native *E. coli* transcription factors that respond to micronutrient levels, and we present strategies for eschewing sensitive detecting equipment for bioluminescent or fluorescent reporters while maintaining unambiguous sensor output. We have demonstrated visible, unambiguous changes in cell color using this circuit, and have investigated multiple configurations of the genetic circuitry needed to produce this color, manipulating strand orientation, gene order, and ribosomal binding sites, ultimately identifying improved designs that allow for more unambiguous coloration in a shorter period of time. We have also investigated the use of additional genetic manipulations to facilitate the tuning of the whole-cell biosensor to respond at the desired micronutrient levels. The resulting biosensor is capable of distinguishing conditions that are high, intermediate, or low in our selected micronutrient, and is currently being engineered for expanded detectable analyte ranges and reduction to practice in a format ready for use in the field.

390397 The Deg-on System: A Mammalian Synthetic Gene Circuit for Sensitive Detection of Proteasomal Activation

Abstract:

Protein homoeostasis in eukaryotic cells relies on the quality control function of the ubiquitin proteasome system (UPS), a complex macromolecular machinery that regulates protein turnover and catalyzes degradation of misfolded or damaged proteins. UPS dysfunction has been linked to the pathogenesis of several diseases, ranging from cancer to cardiovascular and neurodegenerative disorders, implicating the UPS as a molecular target for therapeutic intervention in a diverse range of human diseases. Small molecules and peptides that inhibit the peptidase activities of the proteasome have been widely used to characterize the role of the UPS in maintaining protein homoeostasis and transitioned to the clinic for treatment of cancer and inflammatory diseases. On the other hand, pharmacological agents that increase UPS activity remain largely unexplored, although this mode of regulation has great potential to ameliorate the phenotypes associated with the development of protein deposition diseases, such as Parkinson's and Huntington's. To develop a mammalian cell-based platform to quantify proteasomal degradation that is amenable to high-throughput screening of UPS activators, we created a genetic inverter, the Deg-On system. This orthogonal genetic circuit translates proteasomal degradation of an engineered version of the transcriptional regulator TetR into a fluorescent signal, thereby linking UPS activity to an easily detectable output. The output signal can be finely tuned using tetracycline, enabling calibration of the inverter for the detection of different degrees of activation of UPS function. Predictive modeling was used to improve the sensitivity and dynamic range of the Deg-On system and lead to the design of the 'enhanced Deg-On' (*e*Deg-On) system, a genetic inverter relying on a positive feedback loop that enables self-amplification of engineered TetR. Introducing an artificial positive feedback loop as a genetic amplifier, the eDeg-On system may exhibit bistability, thereby resulting in hysteresis or 'cellular memory' in response to tetracycline dosage and UPS activity. The input–output behavior of the *e*Deg-On system was also confirmed in the context of stable cell lines generated by genetically integrating the relevant parts of the circuits. By providing a cell-based platform to quantify modulation of UPS activity, this *e*Deg-On cell line is used in the screening for compounds that enhance UPS activity, which will open the way to the development of therapeutic strategies to reduce the accumulation of aberrant proteins under conditions of proteotoxic stress.

363076 Novel in Vivo Biosensors for Monitoring of Mammalian Cell Cultures

Abstract:

Mammalian cell cultures are used for production of biopharmaceuticals, *e.g.* monoclonal antibodies. Only mammalian hybridoma cells contain the pathways for antibody production, but due to their multicellular origin the cells have complex nutrient requirements. Cell growth and antibody production is limited by supply of essential nutrients such as glutamine and accumulation of toxic waste products such as lactate. Many attempts have been made at tackling these challenges, *e.g.* by optimising growth media to keep metabolite concentrations at optimal levels. These approaches have been hampered by our ability to monitor relevant cell culture parameters such as metabolite concentration dynamics in real time.

The aim of this study is to develop a solution to this problem using a Synthetic Biology approach. Whole-cell bacterial biosensors for important culture parameters, such as glutamine, leucine, alanine and lactate were designed, built and characterised. The biosensors were designed by applying the principles of standardisation and modularisation to natural metabolite-sensing systems.

These biosensors are also more generally applicable in any experimental context that requires sensing of metabolites. The results of this study highlight the many challenges of applying synthetic biology constructs to complex industrial contexts. Finally, these whole-cell bacterial biosensors have great potential. By linking the detection step to a transcriptional output, the bacterial cells could directly respond to the information by changing culture conditions. This could lead to a low-cost artificial symbiosis system.

387989 Engineering Cell-Based Therapies That Recognize Defined Combinations of Environmental Cues

Abstract:

Technologies enabling bioengineers to construct cell-based therapies that sense and respond to specific combinations of environmental cues may transform our ability to design and implement novel therapeutic strategies that overcome existing barriers to treatment. For example, engineering cell-based immunotherapies that better discriminate between tumors and healthy tissue could improve the safety and efficacy of this promising approach. Implementing such strategies will require synthetic biology "parts" that sense physiologically relevant and exclusively extracellular cues, such as protein mediators of intracellular communication, and relay this sensing event to engineered circuits comprised of synthetic proteins, RNA, and/or DNA. To address this need, we developed the first fully orthogonal cell surface biosensor platform, termed a modular extracellular sensor architecture (MESA). MESA receptors use a generic signal induction mechanism wherein receptor dimerization results in the release of an engineered transcription factor from the plasma membrane. Here, we leverage MESA technology to engineer cell-based devices that perform multiparametric evaluation of extracellular cues using engineered receptors coupled to intracellular gene circuits. We first evaluated the potential for multiplexing MESA receptors – expressing multiple receptors in the same cells. Orthogonal MESA receptors release distinct transcription factors to serve as inputs to a downstream gene circuit. Finally, we describe MESA receptors that utilize a novel class of affinity reagent as a ligand binding domain – the camelid antibody analog termed a "nanobody" – an approach that may ultimately facilitate engineering cell-based therapies that recognize diverse environmental cues. Like many synthetic biology technologies, these platforms provide powerful tools for scientific inquiry and investigation of disease processes as well as novel therapeutic modalities.

389053 Autonomous Control of Metabolism with Synthetic Sensor-Circuits

Abstract:

Micro-organisms have been engineered to manufacture diverse chemical products, including drugs, materials, and fuels; however, production titers are often anti-coupled to the pathway's biosynthesis rate, due to competition for essential metabolites and the accumulation of toxic byproducts. As a result, a key challenge has been to maximize production titers with changing bioreactor conditions and feedstock compositions. To solve this challenge, microbial sensors have been employed to dynamically control pathway flux; for example, using stress-responsive promoters to control isoprenoid biosynthesis (Dahl, Zhang, et. al., 2013) and fatty acid-based products (Zhang, Carothers, & Keasling, 2012). Similar to the chemicals industry, the development of sophisticated metabolic controllers has the potential to adaptively optimize diverse biosynthesis processes, though quantitative design principles have yet to be elucidated.

Here, we developed, experimentally characterized, and computationally modeled sensor-circuit-pathway systems to develop a design approach for selecting the ideal sensor and circuit to control a target pathway. We built sensors that take advantage of the cell's natural stress response to detect growth phase changes, starvation conditions, the presence of inhibitory feedstock components (furfural), and accumulation of toxic byproducts. We experimentally combined sensors with signal amplifier genetic circuits to increase the sensors' control ranges. We then experimentally connected sensor-circuits to a carotenoid biosynthesis pathway that directly competes with cell growth. Sensors, circuits, and the pathway are individually modeled, and combined, to predict how different sensors and circuits control pathway flux dynamics. We propose an initial set of design rules to maximize production titers.

Specifically, our characterized sensors increase protein expression by up to 5-fold, with respect to control, in response to farnesyl diphosphate depletion as a result of activation of an optimally balanced neurosporene biosynthesis pathway with a maximum productivity of 517 ug/gDCW/hour (Farasat et. al., 2014). These sensors are combined with a two-repressor signal amplification circuit to further widen the input-output range by 10-fold. Each component's model has been experimentally validated; sensors are modeled using state-dependent differential equations, circuits are modeled using statistical thermodynamics, and the pathway is modeled using elementary mass action kinetics. We present both graphical and numerical solutions to this classic control problem, applied to biosynthesis pathways, and highlight experimental sensor-circuit-pathway results indicative of typical control behaviors. Our results are the first steps towards developing comprehensive control principles for metabolic systems, while providing re-usable sensors and circuits to dynamically regulate diverse pathways.

Dynamic knockdown of *E. coli* central metabolism for redirecting fluxes of primary metabolites

Irene M. Brockman and Kristala L. J. Prather
Department of Chemical Engineering
Massachusetts Institute of Technology

Synthetic Biology Applications
2014 AIChE Annual Meeting
17 November 2014

Massachusetts Institute of Technology

Research in the Prather lab focuses on conversion of biomass to valuable chemicals

www.spriinc.org

→

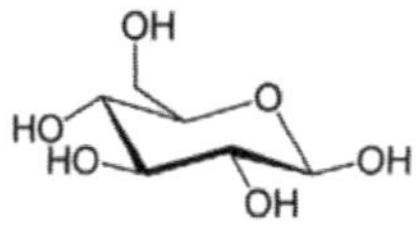

simple sugars

microbial conversion →

biofuels
biopolymers
pharmaceuticals

Massachusetts Institute of Technology

2

Glucaric acid is a "top value added chemical from biomass" with several potential markets

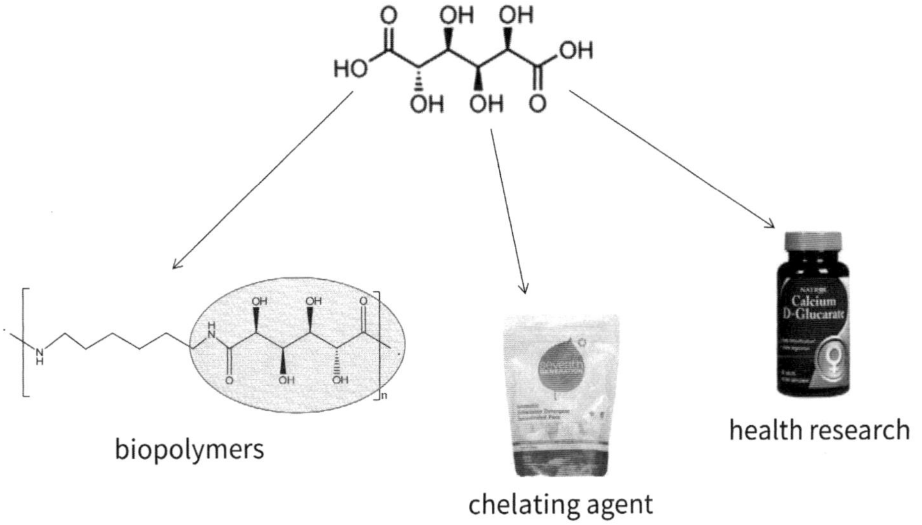

Massachusetts Institute of Technology

Werpy, T. & Petersen, G. USDoE (2004) 3

A pathway for the production of glucaric acid was developed and expressed in *E. coli*

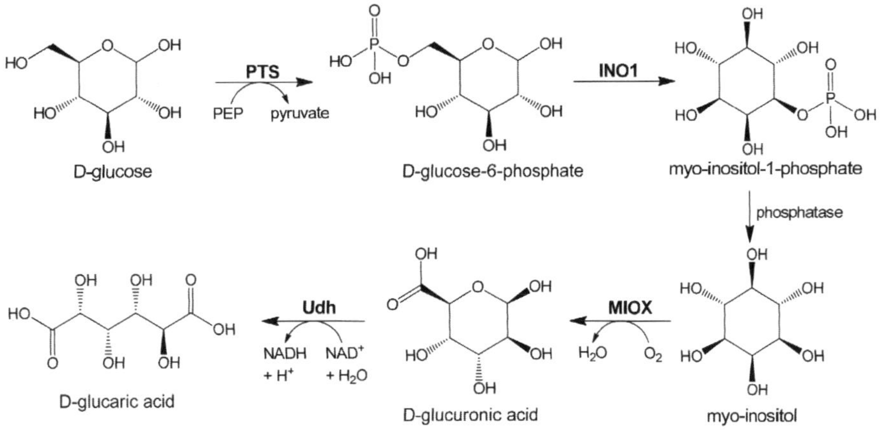

INO1 *S. cerevisiae*
MIOX *M. musculus*
Udh *P. syringae*

- Titers over 1.5 g/L on complex medium (LB) + glucose (10 g/L)
- Low yields (~2%) on glucose minimal medium

Massachusetts Institute of Technology

Moon, T. S., *et. al. Appl Environ Microbiol.* (2009) 4

Gene knockouts in glycolysis compromise cell health

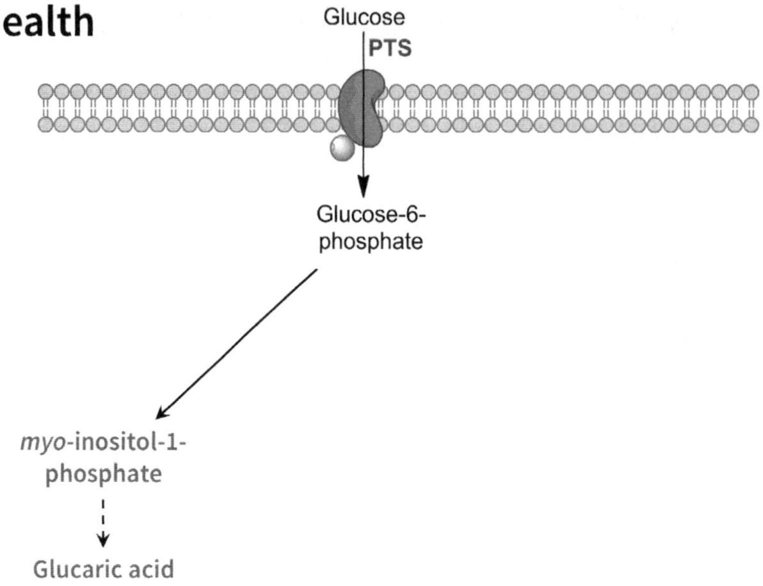

Massachusetts Institute of Technology

5

Project motivation: dynamic control of metabolite flux

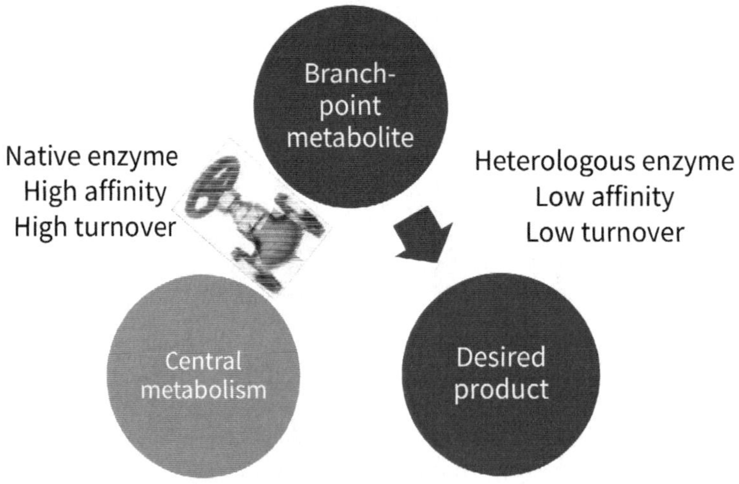

Massachusetts Institute of Technology

6

Strategy for "metabolite valve" development

- Identify enzyme(s) acting as flux control points
- Disrupt any native regulation and tune baseline expression level
- Implement a system to knock down activity of target enzyme in response to fermentation conditions
- Combine the system with expression of heterologous (glucaric acid) pathway

Massachusetts Institute of Technology

7

Phosphofructokinase controls flux into upper glycolysis

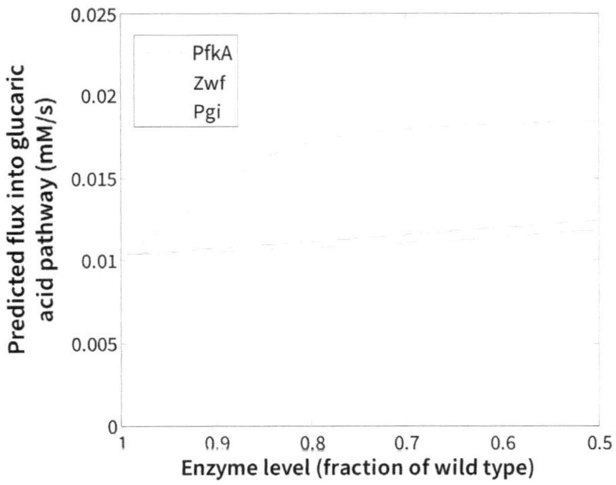

Kinetic model created using literature values for *in vitro* enzyme parameters and metabolite pools

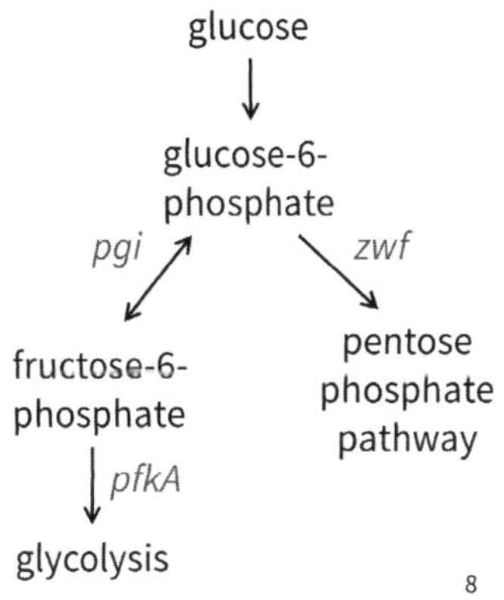

Massachusetts Institute of Technology

8

Native transcriptional control can be disrupted through promoter replacement

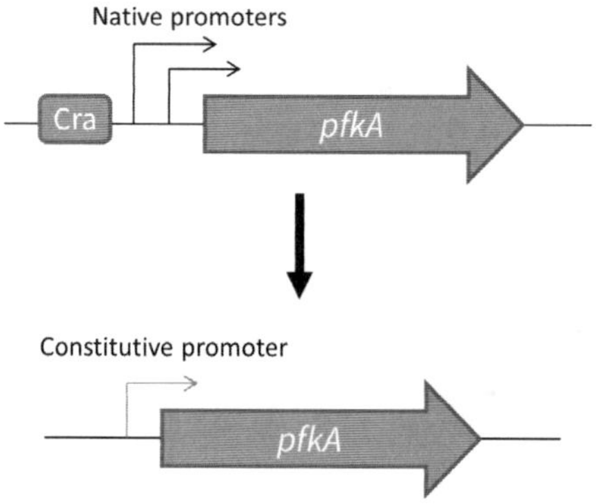

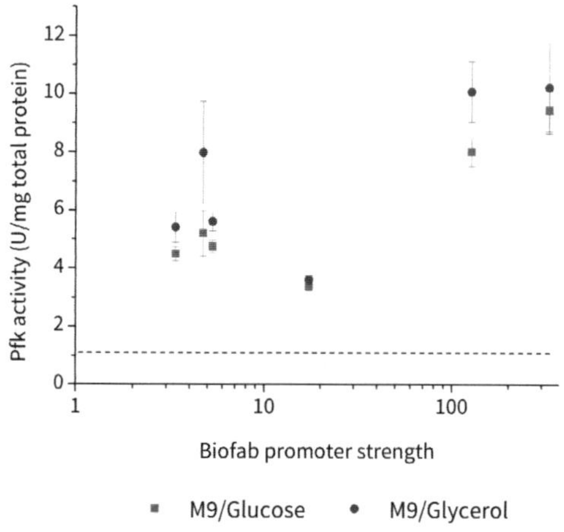

General correlation between promoter strength and activity, but hard to find low strength promoters

MIT Massachusetts Institute of Technology

9

Methods for dynamically controlling enzyme activity

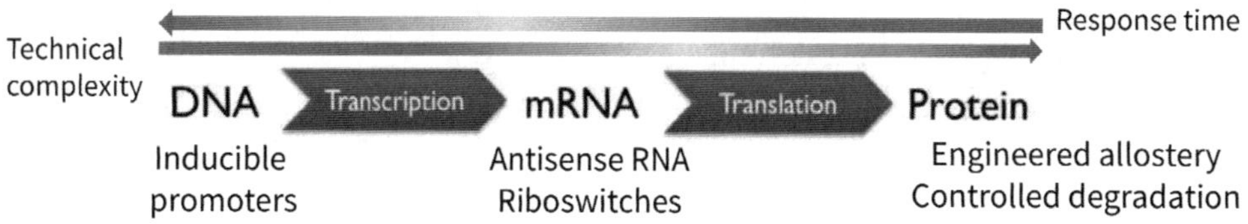

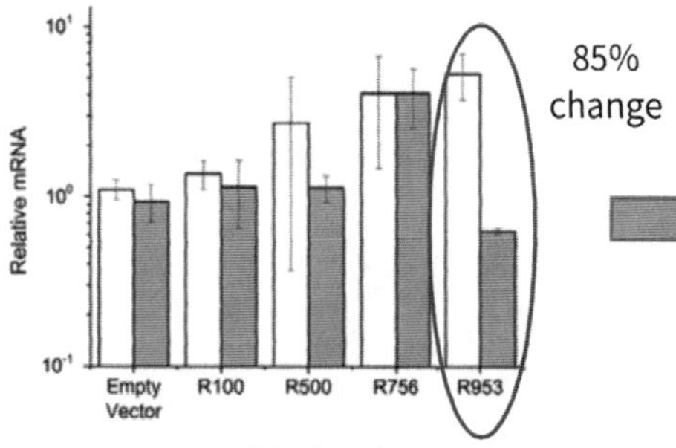

RNA level

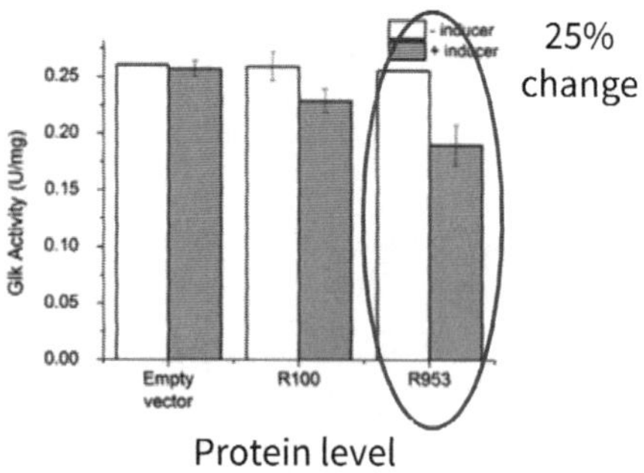

Protein level

MIT Massachusetts Institute of Technology

Solomon, K. *et. al. Metabolic Engineering.* 2012

10

Depletion of essential enzymes slows protein dilution

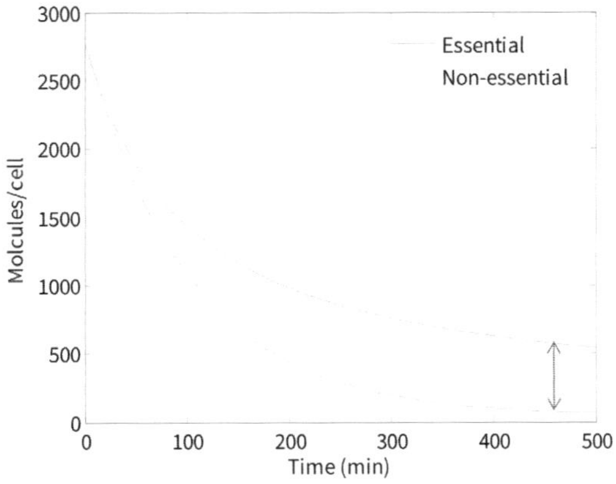

$$\emptyset \xrightarrow{r_1} mss \qquad r_1 = \alpha_m * g$$
$$mss \xrightarrow{r_2} \emptyset \qquad r_2 = (\mu + \beta_m) * mss$$
$$as + mss \xrightarrow{r_3} mds \qquad r_3 = k_h * as * mss$$
$$mds \xrightarrow{r_4} as + mss \qquad r_4 = k_d * mds$$
$$mds \xrightarrow{r_5} \emptyset \qquad r_5 = (\mu + \beta_m) * mds$$
$$mss \xrightarrow{r_6} P + mss \qquad r_6 = \alpha_p * mss$$
$$P \xrightarrow{r_7} \emptyset \qquad r_7 = (\mu + \beta_p) * P$$
$$\emptyset \xrightarrow{r_8} as \qquad r_8 = \alpha_m * g_p$$
$$as \xrightarrow{r_9} \emptyset \qquad r_9 = (\mu + \beta_m) * as$$

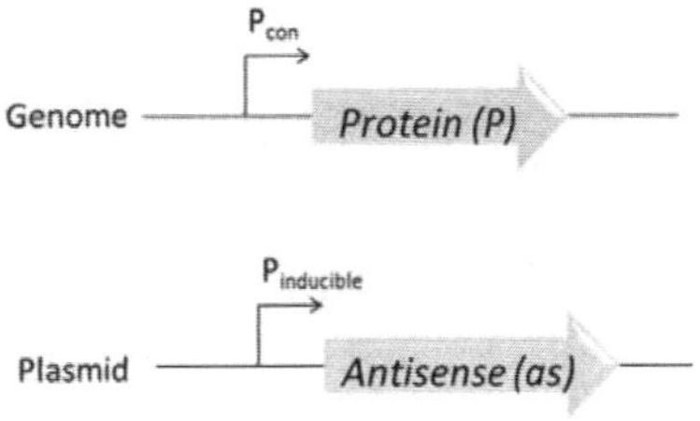

11

Advantages of protein degradation strategies

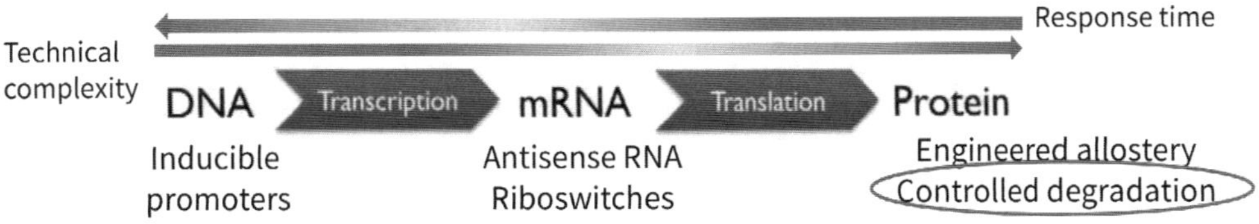

- Transcriptional and post-transcriptional control subject to longer lag times, require dilution by growth to deplete enzyme
- Post-translational control offers more rapid changes in enzyme activity, even under conditions of slow growth
- Drawbacks: complexity, possible wasted resources

12

Adaptor-mediated degradation can be used to achieve dynamic control

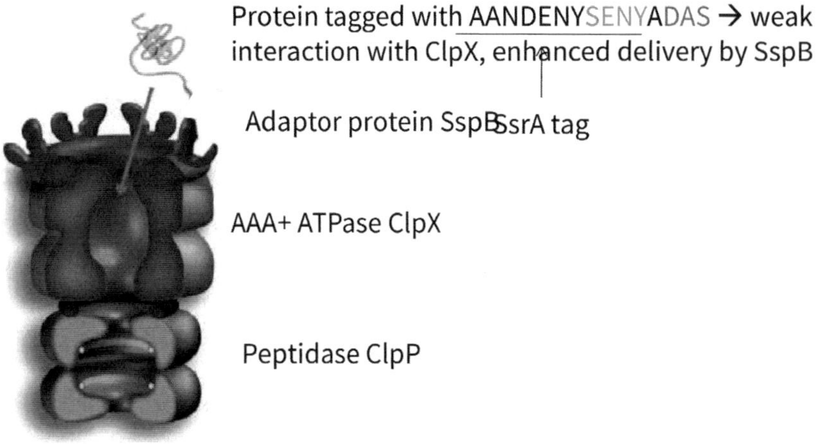

SspB expression can be used to control degradation rate in order to dynamically alter enzyme levels

Massachusetts Institute of Technology

Schmidt, R., *et. al. Research in Microbiology* (2009)
McGinness, K. *et. al. Molecular Cell* (2006)

13

Modifications introduced into *E. coli* MG1655 host strain

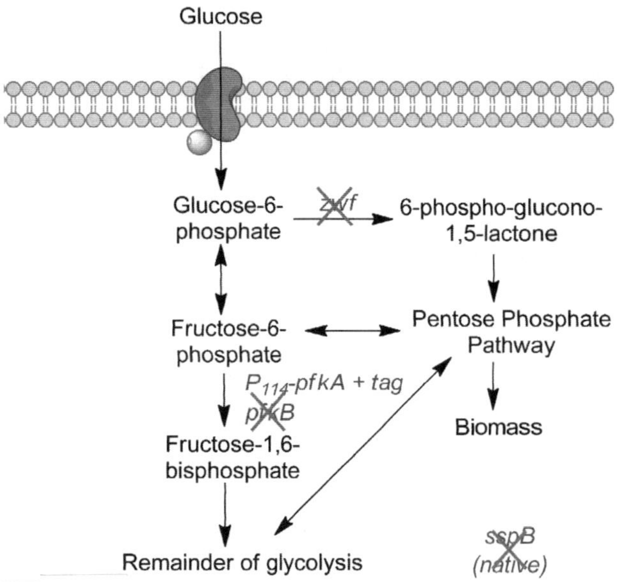

Growth rate still comparable to wild type baseline

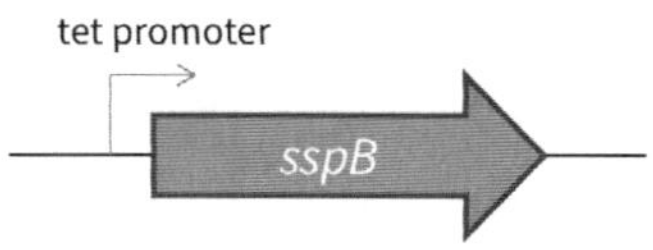

genomic integration

Massachusetts Institute of Technology

14

Typical experimental setup

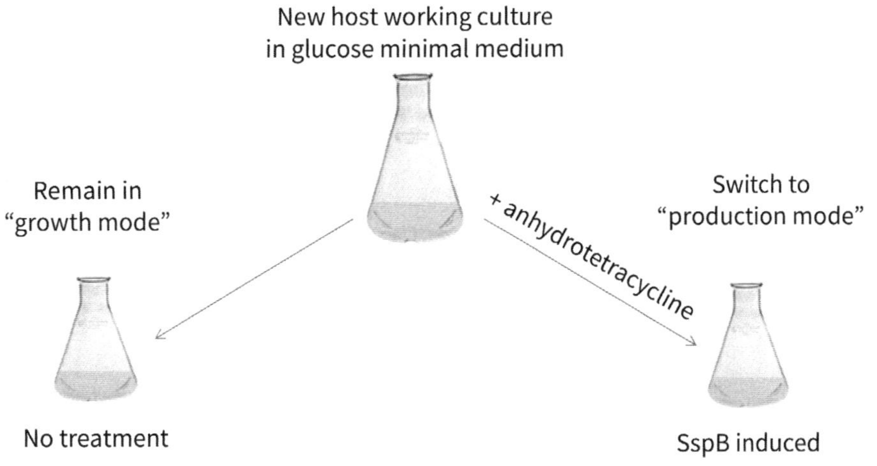

Control strain: wild type promoter and *pfkA* coding sequence

MIT Massachusetts Institute of Technology

15

Induction of SspB results in a rapid decline in Pfk activity levels

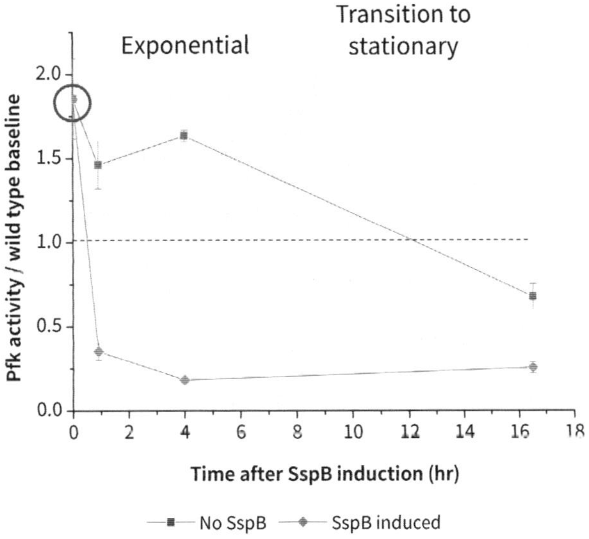

Modified MOPS minimal medium + 1% glucose, 30° C

MIT Massachusetts Institute of Technology

16

Can achieve a variety of growth rates on glucose through induction of SspB

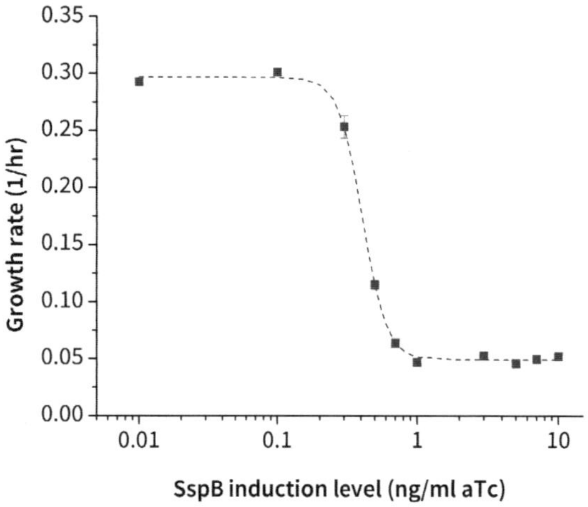

Modified MOPS minimal + 1% glucose, 30° C

Control strains show constant growth with SspB induction

Production mode shows increased sugar phosphate pool

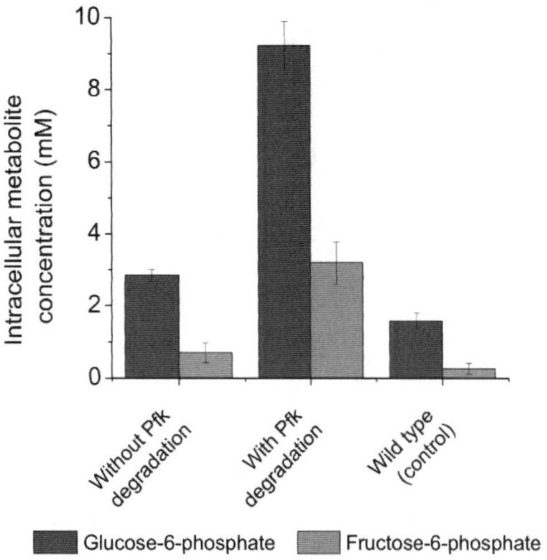

Modified MOPS minimal + 1% glucose, 30° C

Cell mass collected by rapid filtration and intracellular metabolites extracted into boiling 75% ethanol

G6P / F6P ratio is consistent with expected equilibrium

Coupling the "valve" strain with the glucaric acid pathway

- Initial testing: expression of INO1 only and quantification of *myo*-inositol via HPLC
- Induction of SspB controlled by aTc addition at various times from inoculation until stationary phase

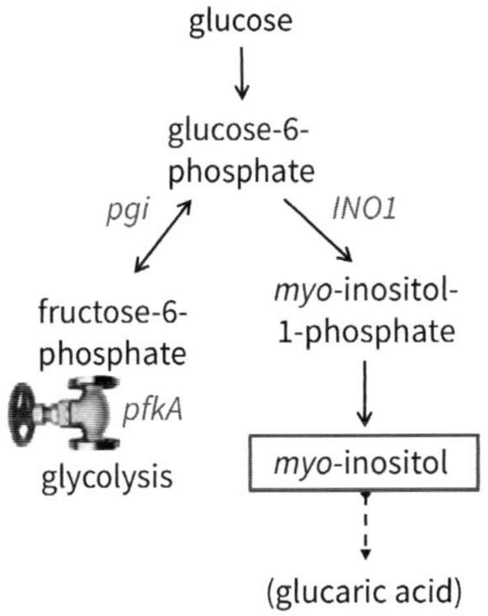

Massachusetts Institute of Technology

19

Growth profile shows expected growth arrest at SspB induction

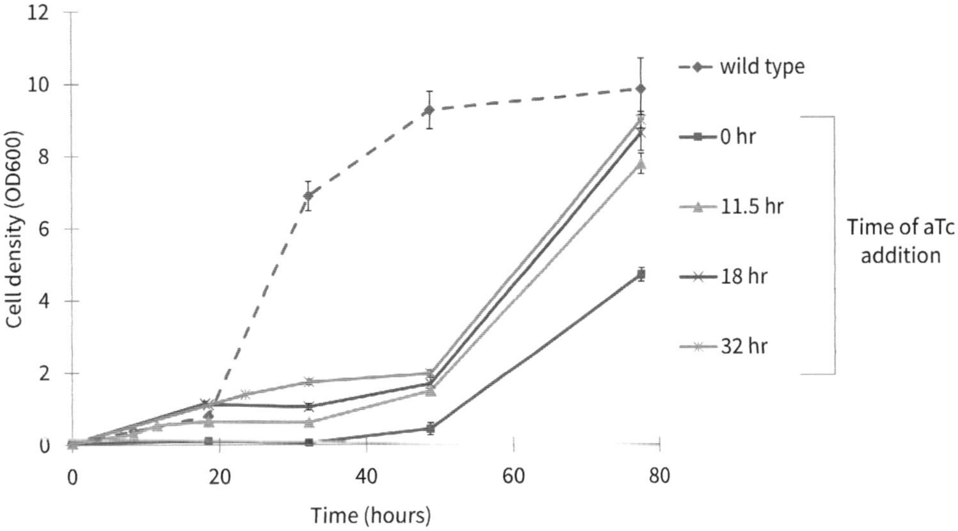

Massachusetts Institute of Technology

20

Dynamic Pfk knockdown results in two-fold improvement in *myo*-inositol yield and titers

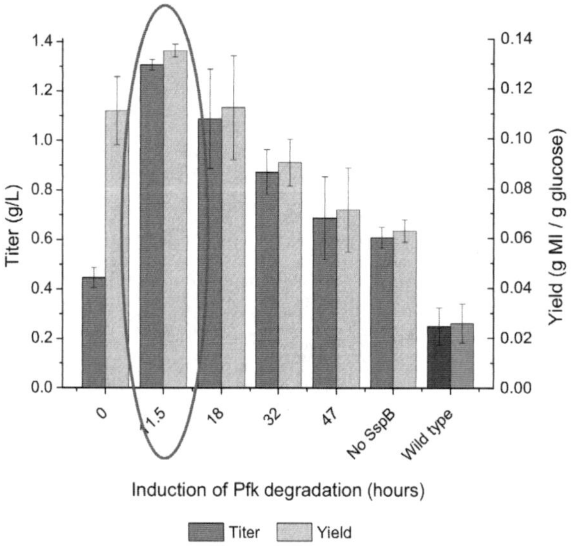

Modified MOPS minimal medium + 1% glucose

78 hour culture at 30° C

Brockman, I.M. and Prather K.L.J, submitted 21

Project summary and future work

- Developed a general workflow for implementation of the metabolite valve concept in *E. coli*

- Successfully implemented a system for redirecting flux in upper glycolysis that can be used to control cell physiology and increase *myo*-inositol yields and titers

- Ongoing work includes
  - Integration with full glucaric acid pathway
  - Development of transcriptional controls for autonomous switching to production mode
  - Exploration of optimal switching time based on metabolic models

22

Acknowledgements

- Prather Lab
  - Prof. Kristala L.J. Prather
  - Dr. Kevin Solomon
  - Dr. Eric Shiue
  - Apoorv Gupta

- Funding sources
  - NIH Biotechnology Trainee Program
  - NSF

Massachusetts Institute of Technology

23

Development of Gene-Regulatory Control Tools for Antibiotic Technologies

Brittany Forkus and Yiannis Kaznessis
Department of Chemical Engineering and Material Science
University of Minnesota

Antibiotic resistant bacterial infections have recently emerged as an escalating threat to human health. Bacterial pathogens are evolving a multitude of antibacterial evasion mechanisms at a far greater rate than the development of new drugs. Resistant microbes easily transfer across international borders, human-animal ecosystems, and bacterial kingdoms, greatly diminishing the efficacy of existing first-line antibiotics. Gram-negative pathogens, such as *Salmonella* and *E.coli,* are of particular concern as they are the major causative agents of human foodborne illnesses. However, due to these strains' superior abilities to rapidly acquire and communicate their resistance determinants, there is a limited supply of remaining treatment options. In this work, synthetic biology techniques are applied to develop probiotic 'cellbots' capable of alternatively treating resistant foodborne infections.

Bacteriocins are toxic proteins natively produced by *E.coli* which demonstrate remarkable antibacterial against diarrheagenic pathogens. However, their potential clinical success is limited by their susceptibility to proteolysis and poor pharmacokinetic behavior. In this work, probiotic *E.coli Nissle 1917* is being re-engineered to overcome these limitations by safely producing and delivering substantial doses of bacteriocins at the localized site of infection. A fundamental challenge with drug-delivery systems is the requirement for well-characterized and tightly-controlled protein production. The envisioned cellbots need to maintain a pathogen-killing capacity while minimizing host systemic effects. Synthetic gene-regulatory networks are attractive frameworks for they may give rise to complex biological phenotypes which cannot be otherwise achieved, potentiating broader dynamical control over gene expression.

In this work, an inducible bacterial switch, termed ProTeOn+, has been developed which elicits minimal basal protein expression levels while demonstrating robust and sensitive control over green fluorescent protein (GFP) production. The prokaryotic device was developed by stitching together pieces of the bacterial tetracycline, luminescence, and lactose operons to create a tetracycline-inducible phenotype. A combination of high-resolution fluorescence microscopy and flow cytometry techniques have been utilized to yield insight at the population and single-cell level of this inherently stochastic cellular process. Fluorescence single-cell tracking experiments are being developed to observe the dynamic protein production trajectories of the microbes with the intention of developing predictive stochastic simulations.

ProTeOn+ is currently being modified to express the antimicrobial peptides, A3APO, Alysteserin, and Microcin J25. In previous work, these peptides have shown significant antibacterial activity against many clinically-relevant strains of *Salmonella* and *E.coli.* The envisioned probiotic cellbots will use the ProTeOn+ production platform to deliver these AMPs to the intestinal environment in a spatiotemporally controlled manner. The effective development of the recombinant probiotic strains detailed in this work would be an asset to antibacterial treatments and a demonstration of the capabilities of synthetic biology.

390155 Parallel Mapping of Cryptic and Multidrug Resistance Alleles in Bacteria

Abstract:

Antibiotic resistance in microbes constitutes an impending crisis in clinical, agricultural, and industrial settings. Here we use genome engineering tools enabled by synthetic biology to examine factors underlying the development of antibiotic resistance in bacteria. A barcoded, genome-scale library of strains simulating overexpression or knockout of over 4,000 genes in *E. coli* (the TRMR library) was challenged in parallel at sub-inhibitory concentrations with eight different clinically-relevant antibiotics. The selected populations were rapidly characterized by next-generation sequencing and the approximately 200,000 individual fitness measurements were analyzed in the QIIME (Quantitative Insights into Microbial Ecology) bioinformatics pipeline. We discovered that the most prevalent alleles conferred resistance to multiple antibiotics (multi-drug resistant, or MDR alleles) and several of these isolated alleles were previously unknown or uncharacterized. Interestingly, chemically similar antibiotics or antibiotics with similar mechanisms of action did not produce similar allelic responses on the genome level. Also, in spite of the prevalence of MDR alleles, the depth of characterization enabled by next-generation sequencing allowed the use supervised learning to determine a genomic "fingerprint" for each antibiotic of interest. The same methodologies of rapid selection, high-throughput sequencing, and bioinformatic analysis are broadly applicable to experiments on chemical tolerance for any inhibitory chemical, from antibiotics to toxic metabolites to next-generation biofuels.

386910 An Improved Genome-Scale Metabolic Network Model for Scheffersomyces Stipitis

Abstract:

Due to its important role in the production of lignocellulosic ethanol, a second generation biofuel, *Scheffersomyces stipitis* has drawn tremendous research interest in the last few decades. This is because S. stipitis has the highest native capacity to ferment xylose into ethanol and how to efficiently utilize the pentose sugars in lignocellulosic hydrolysate remains one of the major barriers in producing cost-effective lignocellulosic ethanol through biological conversion. Although S. stipitis can produce ethanol at a yield close to the theoretical maximum under microaerobic condition, its xylose consumption rate is only half of that under fully respiratory conditions. In addition, low ethanol tolerance and no growth under anaerobic condition also limit its direct utilization in industrial applications. To address these limitations, attempts have been made to improve xylose pick up rate of S. stipitis under microaerobic condition, as well as introducing xylose fermentation pathways possessed by S. stipitis to engineered Saccharomyces cerevisiae. However, these attempts have not been entirely successful: for engineered S. stipitis strains.

Because of the complexity involved in the two yeast strains, a better understanding on cellular metabolism of *S. stipitis* would be helpful for improving both S. stipitis and engineered S. cerevisiae. Because genome-scale metabolic models connect the genotype with phenotype, they provide a holistic view of the cellular metabolism. Once validated, genome-scale models provide a platform to effectively interrogate cellular metabolism, such as characterizing metabolic resource allocation, predicting phenotype, designing experiments to verify model predictions, as well as designing mutant strains with desired properties. More importantly, genome-scale models allow systematic assessment of how a (genetic or environmental) perturbation would affect the organism as a whole.

The complete genome of S. stipitis has been sequenced, which provides the foundation for genome-scale metabolic network reconstruction. Recently, two genome-scale models have been published, iSS844[1] and iBB814[2], which represents a significant step forward in gaining systems understanding of cellular metabolism of S. stipitis. Both models have fair amount of experimental validations and both performed well in matching the model predictions with some experimental measurements. However, both models have certain limitations. For example, neither of them predicts the production of xylitol under any condition, aerobic, microaerobic or anaerobic. It is well-known that xylitol is a key byproduct produced by *S. Stipitis* under microaerobic condition due the redox imbalance caused by xylose fermentation step (XR and XDH). Therefore, being able to predict xylitol production under microaerobic condition is very important for a model to identify candidates for balancing cellular redox potential.

In our previous work, we have conducted comprehensive evaluation of the two genome-scale models. The evaluation consists of three steps: first is the manual comparison of the reactions included in both models; second is the conduction of *in silico* experiments to examine model's global behavior as well as comparing the model predictions with additional experimental results reported in literature; Final step is the application of the system identification framework we developed to extract biological knowledge from carefully designed *in silico* experiments, and to examine whether the extracted knowledge agrees with existing understandings. Our finds indicate that even though model iBB814 show worse agreement with addition experimental results, it agrees better with existing biological knowledge, and therefore was chosen as the base model for further improvement.

In this work, guided by the findings obtained through the system identification framework, we developed a modified genome-scale mode for S. stipitis. The modified model not only shows better agreement with various experimental results, but also fully agrees with existing knowledge on how different pathways (glycolysis and pentose phosphate pathway) would respond to increased substrate pickup rates (such as oxygen pickup rate). In addition, the improved model predicts xylitol production under oxygen-limited conditions, and our analysis show that the purine metabolism in conjunction with the pentose phosphate pathway is critical for xylitol production. The modified genome-scale model allows us to examine the details of cofactor imbalance under oxygen-limited conditions, as well as examine the effect of different redox ratio of xylose reductase and xylitol dehydrogenase on ethanol production. Based on these analyses, metabolic engineering strategies were proposed for enhanced ethanol production.

388925 A Kinetic Model of Escherichia coli Core Metabolism Satisfying Multiple Sets of Mutant Flux Data

Abstract:

In contrast to stoichiometric-based models, the development of large-scale kinetic models of metabolism has been hindered by the challenge of identifying kinetic parameter values and kinetic rate laws applicable to a wide range of environmental and/or genetic perturbations. The recently introduced ensemble modeling (EM) procedure was developed towards achieving this goal through successively reducing the parameter space by using phenotypic data such as experimental flux measurements and decomposing the reactions into the elementary mechanism. Here, we developed an optimization-driven parameter estimation method and combined it with EM approach to construct a large-scale kinetic models of E. coli metabolism satisfying the fluxomic data for wild-type and seven mutant strains. This model encompasses 138 reactions and 93 metabolites accounting for glycolysis/gluconeogenesis, pentose phosphate pathway, TCA cycle, major pyruvate metabolism, anaplerotic reactions, and a number of reactions in other parts of the metabolism and accounts for 60 regulatory reactions that are mainly included the central metabolism interactions. Parameterization is performed using a formal optimization approach that minimizes the discrepancies between model predictions and flux measurements. The predicted fluxes by the model are within the uncertainty range of experimental flux data for 78% of the reactions (with measured fluxes) for both the reference (wild-type) and seven mutant strains. The remaining flux predictions fall within three standard deviations of measured values. Converting the EM-based parameters into a Michaelis-Menten equivalent formalism revealed that 80% of K_m and k_{cat} parameters are within one order of magnitude of literature available values. The predicted metabolite concentrations by the model are also within uncertainty ranges of metabolomic data for 68% of the metabolites. A leave-one-out cross-validation test to evaluate the flux prediction performance of the model showed that metabolic fluxes for the mutants located in the proximity of mutations used for training the model are predicted more accurately.

The constructed model and parameterization procedure provides the means for the construction of both larger models and models with more narrowly distributed parameter values as new metabolomics/fluxomic data sets are becoming available for E. coli and other well studied organisms.

390673 Synchronization of Cellular Skeleton Determines the Nature of Motions of Cancerous Vs. Non-Cancerous Cells

Abstract:

Protrusions/retractions arise from forces exerted by cell cytoskeleton components to the front and back of the cells. Here, we study if overall motility strategy (diffusive versus Lévy walks) can be emerged from the front/back dynamics and weather synchronization between the two affect the overall cell movement. We have shown that a series of persistent non-Markovian infinitesimal moves could result in a power law distribution of the persistence length characterizing a Lévy walk distribution. To consider the protrusions/retractions of the cell with a primitive model represented by two connected moving points. These front and back compartment of the model were allowed to move with or without synchronization considering limitation on the overall cell size. The statistics of this unidirectional excursions were then collected to characterizing probability distributions type of motion. The model showed that the synchronization strategy affects overall cell motility pattern.

Minimal Reaction Network for Oscillations in the MAPK Signalling Cascade

O. Hadač, M. Přibyl, I. Schreiber*

DEPARTMENT OF CHEMICAL ENGINEERING

INSTITUTE OF CHEMICAL TECHNOLOGY PRAGUE

2014 AIChE Annual Meeting, Atlanta, Nov 16-21, 2014

Outline

- MAPK signalling cascades
  - Mitogen Activated Protein Kinases (MAPKs)
  - MAPK signalling cascades
  - Huang-Ferrell model of MAPK cascade
  - Stage 1 & 2 subsystems of the H-F model
- Reaction network theory
  - Stoichiometric network analysis
  - Reaction networks of the subsystems
  - Stability and reduction of the networks
  - Core bistable subnetwork
  - Extension to a core oscillatory subnetwork
- Conclusions

Mitogen Activated Protein Kinases (MAPKs)

- ❑ Ser/Thr protein kinases
  - widely conserved in eukaryotes
  - activated by phosphorylation on residues
  - deactivated by dephosphorylation on residues
- ❑ MAPKs phosphorylate
  - enzymes
  - transcription factors
  - cytoskeletal proteins
- ❑ MAPKs are involved in three-stage signallig cascades

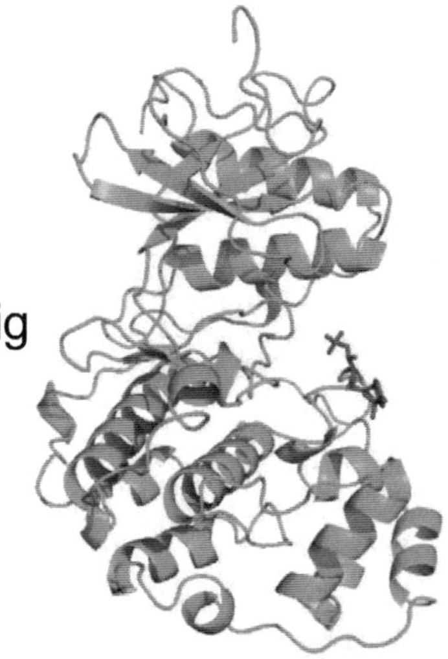

X-ray structure of ERK2 (MAPK) in its active form. Phosphorylated residues are shown in red.

http://en.wikipedia.org/wiki/MAPK

MAPK signalling cascades

- ❑ organized hierarchically into three-tiered modules
- ❑ regulate
  - proliferation
  - gene expression
  - differentiation
  - mitosis
  - cell survival
  - apoptosis
- ❑ Eukaryotic cells
  - 12 MAPKKKs
  - 7 MAPKKs
  - 8 MAPKs

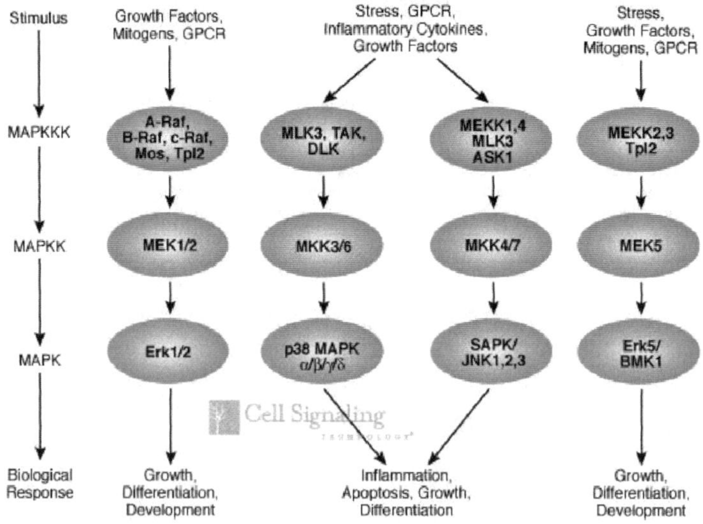

http://www.cellsignal.com/

Huang-Ferrell model of the MAPK cascade

Huang & Ferrell, PNAS **93**, 19, 1996

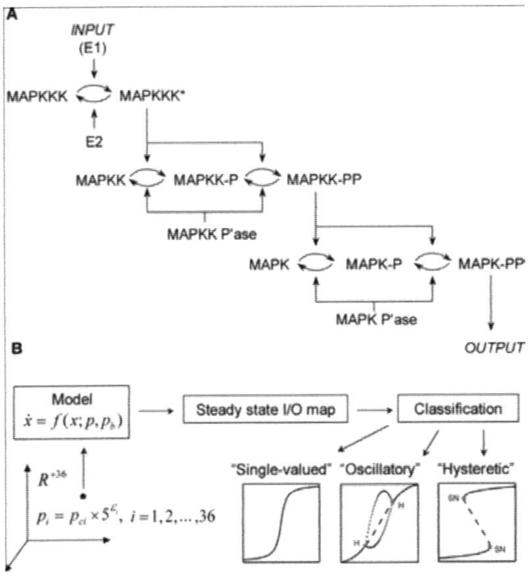

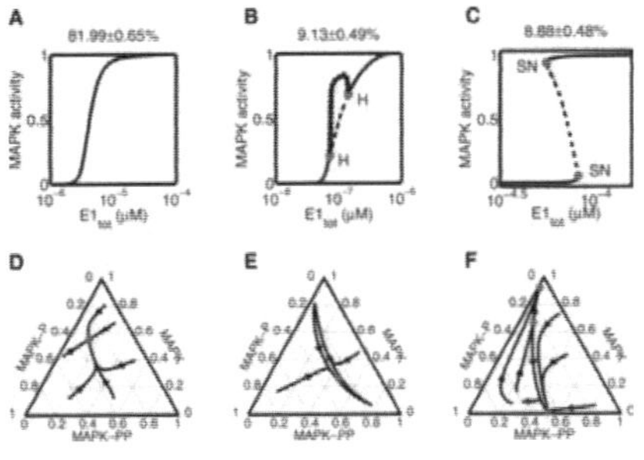

System	D_p [a]	D_v [b]	Single-Valued (Percent)	Oscillatory (Percent)	Hysteretic (Percent)
1 + 2 + 3 [c]	36	15	81.99 ± 0.65	9.13 ± 0.49	8.88 ± 0.48
2 + 3	28	12	73.35 ± 0.75	7.52 ± 0.45	19.13 ± 0.67
1 + 2	22	9	93.65 ± 0.41	1.48 ± 0.20	4.87 ± 0.36
1	8	3	100	0	0
2	14	6	83.88 ± 0.62	0	16.12 ± 0.62
3	14	6	85.75 ± 0.59	0	14.25 ± 0.59

[a] D_p is the total number of free parameters in the corresponding model.
[b] D_v is the number of degrees of freedom of the ODE model of the corresponding biochemical network.
[c] Numbers 1, 2, and 3 refer to the stages of the MAPK cascade corresponding to the activation of MAPKKK, double-phosphorylation of MAPKK, and MAPK, respectively.
doi:10.1371/journal.pcbi.0030184.t001

Qiao L, Nachbar RB, Kevrekidis IG, Shvartsman SY (2007) Bistability and Oscillations in the Huang-Ferrell Model of MAPK Signaling. PLoS Comput Biol 3(9): e184.

Stage 2 of the H-F model

- ❑ What is the core of bistability in the H-F model?

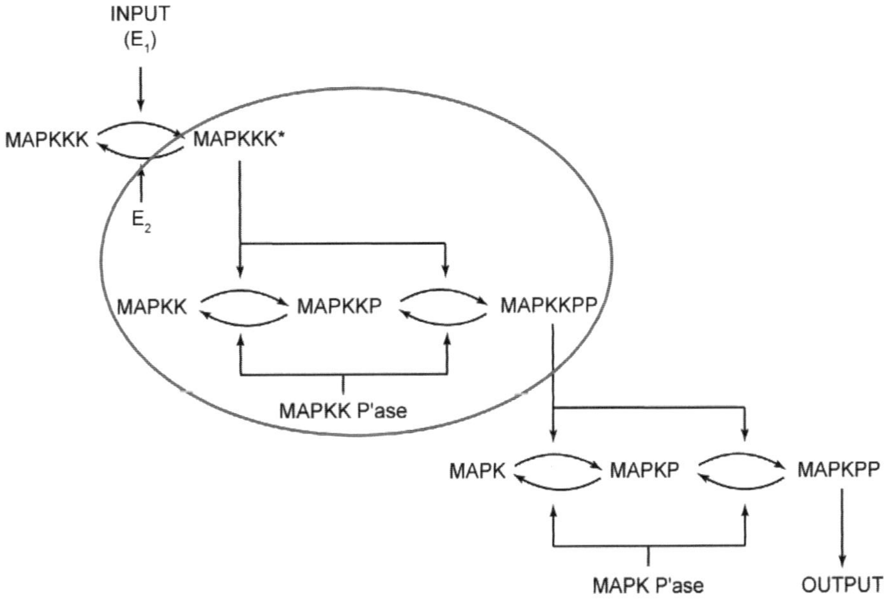

Stage 1& 2 of the H-F model

- ❑ What is the core of oscillations in the Huang-Ferrell model?

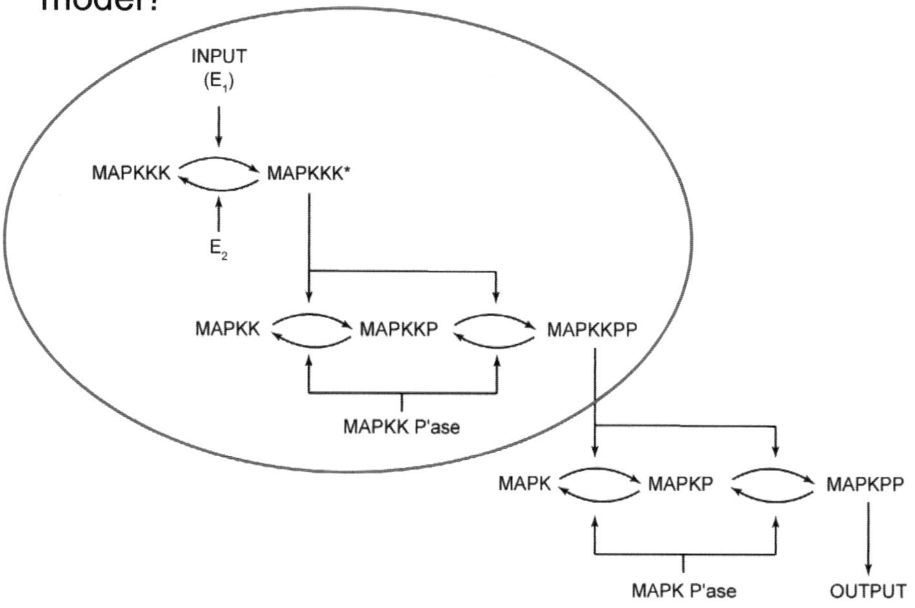

Stoichiometric network analysis (SNA)

- ❑ A graph oriented approach to stability of reaction networks

  B. L. Clarke, Adv. Chem. Phys. 43, 1 (1980)
  J. Ross, I. Schreiber and M.O. Vlad, Determination of Complex Reaction Mechanisms, Oxford University Press, New York, 2006

  - decomposition of reaction networks into irreducible (elementary or extreme) subnetworks
  - reduction of model mechanisms
  - linear stability analysis and graphical interpretation
  - identification of positive and negative feedback loops
  - identification of sources of instabilities (autocatalysis)
  - combination of autocatalysis and negative feedback leads to bistability (O.Hadač, I.Schreiber, JCP, 2013) or oscillations
  - systemic verification of sources of instabilities in reaction mechanisms (O.Hadač, I.Schreiber, PCCP, 2011)
- ❑ other approaches – balanced complexes M. Feinberg, CES 1987,1988

Reaction network of the bistable subsystem

No.	Reaction	Reaction rate
(1)	$B + A^* \to A^*B$	$v_1 = k_1[A^*][B]$
(2)	$A^*B \to B + A^*$	$v_2 = k_2[A^*B]$
(3)	$A^*B \to A^* + B_1$	$v_3 = k_3[A^*B]$
(4)	$B_1 + A^* \to A^*B_1$	$v_4 = k_4[A^*][B_1]$
(5)	$A^*B_1 \to B_1 + A^*$	$v_5 = k_5[A^*B_1]$
(6)	$A^*B_1 \to B_2 + A^*$	$v_6 = k_6[A^*B_1]$
(7)	$B_2 + E_3 \to E_3B_2$	$v_7 = k_7[E_3][B_2]$
(8)	$E_3B_2 \to B_2 + E_3$	$v_8 = k_8[E_3B_2]$
(9)	$E_3B_2 \to B_1 + E_3$	$v_9 = k_9[E_3B_2]$
(10)	$B_1 + E_3 \to E_3B_1$	$v_{10} = k_{10}[E_3][B_1]$
(11)	$E_3B_1 \to B_1 + E_3$	$v_{11} = k_{11}[E_3B_1]$
(12)	$E_3B_1 \to B + E_3$	$v_{12} = k_{12}[E_3B_1]$

Michaelis-Menten kinetics

Symbol	Chemical species
A^*	$MAPKKK^*$
B	MAPKK
B_1	MAPKKP
B_2	MAPKKPP
E_3	E_3
A^*B	$MAPKKK^*MAPKK$
A^*B_1	$MAPKKK^*MAPKKP$
E_3B_1	$E_3MAPKKP$
E_3B_2	$E_3MAPKKPP$

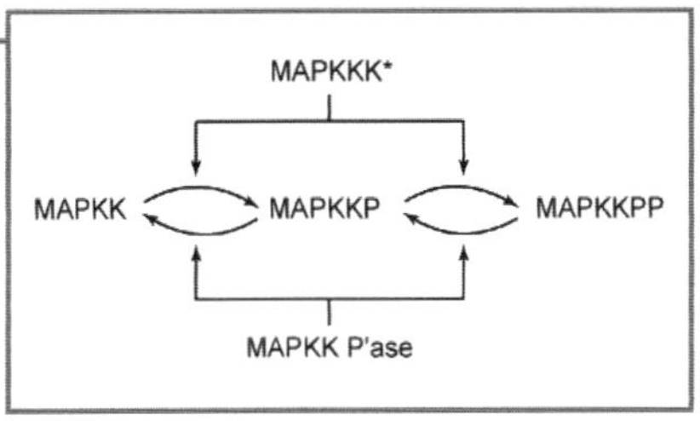

$$\frac{d[A^*]}{dt} = -v_1 + v_2 + v_3 - v_4 + v_5 + v_6,$$
$$\frac{d[B]}{dt} = -v_1 + v_2 + v_{12},$$
$$\frac{d[B_1]}{dt} = v_3 - v_4 + v_5 + v_9 - v_{10} + v_{11},$$
$$\frac{d[B_2]}{dt} = v_6 - v_7 + v_8,$$
$$\frac{d[E_3]}{dt} = -v_7 + v_8 + v_9 - v_{10} + v_{11} + v_{12},$$
$$\frac{d[A^*B]}{dt} = v_1 - v_2 - v_3,$$
$$\frac{d[A^*B_1]}{dt} = v_4 - v_5 - v_6,$$
$$\frac{d[E_3B_1]}{dt} = v_{10} - v_{11} - v_{12},$$
$$\frac{d[E_3B_2]}{dt} = -v_7 + v_8 + v_9.$$

kinetic equation

$$\frac{d\mathbf{x}}{dt} = \dot{\mathbf{x}} = \underline{\nu}\,\mathbf{v}(\mathbf{x}, \mathbf{k})$$

$$A_{tot} = [A^*] + [A^*B] + [A^*B_1],$$
$$B_{tot} = [B] + [B_1] + [B_2] + [A^*B] + [A^*B_1] + [E_3B_1] + [E_3B_2],$$
$$[E_3]_{tot} = [E_3] + [E_3B_1] + [E_3B_2].$$

conservation constraints

Network diagram

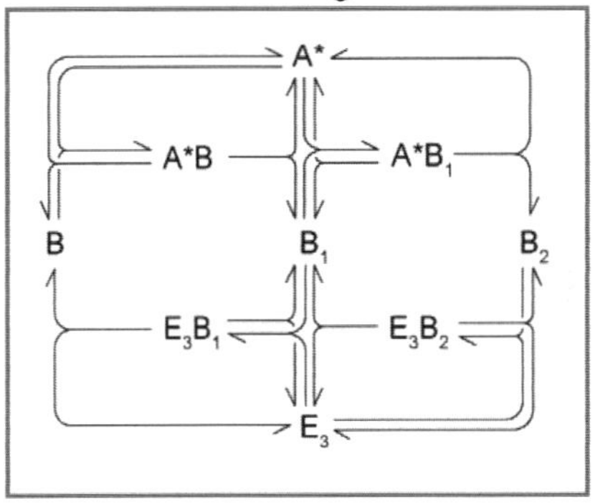

Stability of the reaction network

- Network decomposition
  - find simplest steady state solutions $\mathbf{v}^0$ of

$$\dot{\mathbf{x}} = \underline{\nu}\,\mathbf{v}^0 = \mathbf{0}$$

which represent elementary subnetworks $\mathbf{E}$

Any solution is then

$$\mathbf{v}^0 = \mathbf{E}\,\mathbf{j}$$

$$\mathbf{E} = \underbrace{[\mathbf{E}_1\ \mathbf{E}_2\ \mathbf{E}_3\ \mathbf{E}_4\ \mathbf{E}_5\ \mathbf{E}_6]}_{\text{elementary subnetworks (ES)}} = \begin{bmatrix} 1&0&0&0&1&0\\ 1&0&0&0&0&0\\ 0&0&0&0&1&0\\ 0&1&0&0&0&1\\ 0&1&0&0&0&0\\ 0&0&0&0&0&1\\ 0&0&1&0&0&1\\ 0&0&1&0&0&0\\ 0&0&0&0&0&1\\ 0&0&0&1&1&0\\ 0&0&0&1&0&0\\ 0&0&0&0&1&0 \end{bmatrix}$$

- Linear stability analysis
  - examine eigenvalues of Jacobi matrix

$$\mathbf{J}(\mathbf{h}, \mathbf{j}) = \underline{\nu}\,\mathrm{diag}(\mathbf{E}\,\mathbf{j})\,\underline{\boldsymbol{\kappa}}^{\mathrm{T}}\,\mathrm{diag}(\mathbf{h})$$

effective order of the *j*-th reaction with respect to the *i*-th species

$$\kappa_{ij} \equiv \frac{\partial \ln v_j(\mathbf{x}, \mathbf{k})}{\partial \ln x_i}$$

reciprocal steady state concentration of i-th species

$$h_i = \frac{1}{x_i^0}$$

  - **h** and **j** are convex parameters
  - $E_1 - E_6$ are stable ES
  - subnetwork E_5 & E_6 may become unstable

Reduction of the network

Original network of the stage 2 subsystem

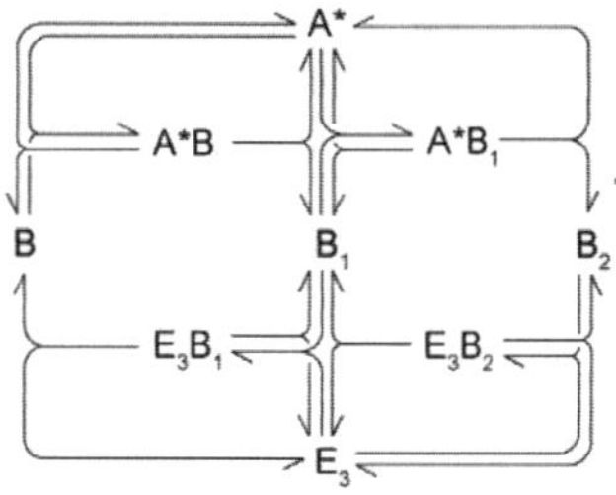

SNA decomposition

Subnetwork E_5 & E_6

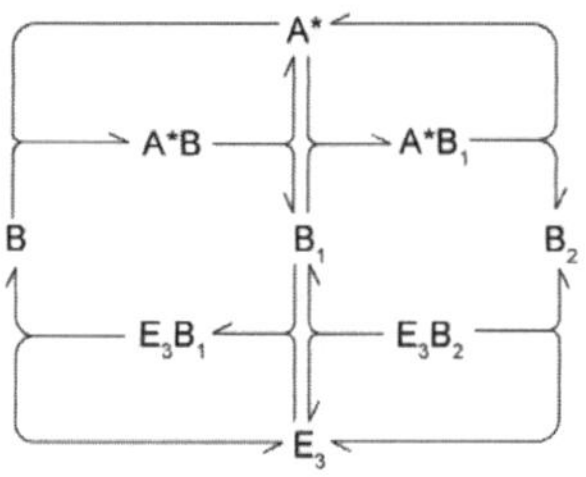

Quasi steady state approx. for A^*B_1 and E_3B_1

Symmetric bistable network

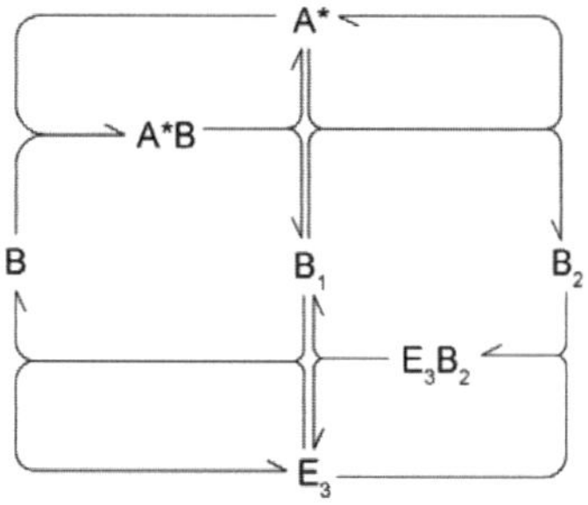

QSSA for E_3B_2 and conc. E_3 is very high

Minimal bistable network

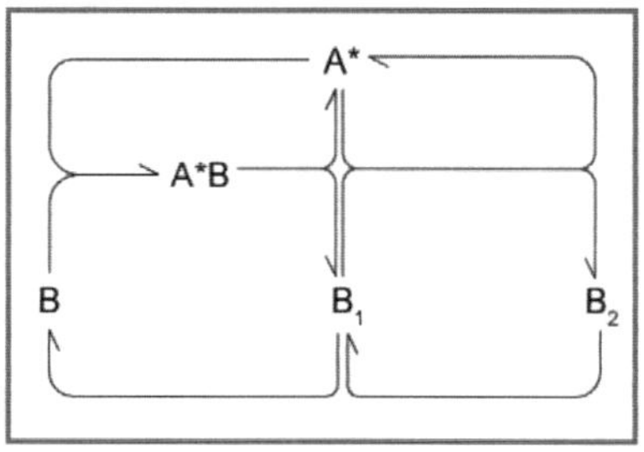

Direct analysis of the minimal bistable network

Minimal bistable network

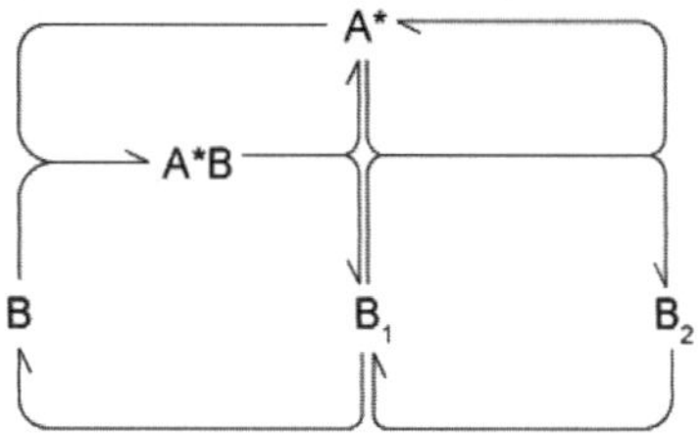

Model equations

$$\frac{dx_1}{dt} = k_2x_5 - k_1x_1x_2,$$
$$\frac{dx_2}{dt} = -k_1x_1x_2 + k_5x_3,$$
$$\frac{dx_3}{dt} = k_2x_5 - k_3x_1x_3 + k_4x_4 - k_5x_3,$$
$$\frac{dx_4}{dt} = k_3x_1x_3 - k_4x_4,$$
$$\frac{dx_5}{dt} = k_1x_1x_2 - k_2x_5.$$
$$A_{tot} = x_1 + x_5\,,$$
$$B_{tot} = x_2 + x_3 + x_4 + x_5\,.$$

Analytical steady state solution

$$k_1k_2k_3x_1^3 + (k_1k_4k_5 + k_1k_2k_4 - k_1k_2k_3A_{tot})x_1^2 + (k_2k_4k_5 - k_1k_4A_{tot}(k_2 + k_5) + k_1k_4k_5B_{tot})x_1 - k_2k_4k_5A_{tot} = 0\,,$$

$$x_3 = k_2(A_{tot} - x_1)/k_5\,,$$
$$x_4 = k_2k_3x_1(A_{tot} - x_1)/(k_4k_5)\,.$$

Bifurcation diagram

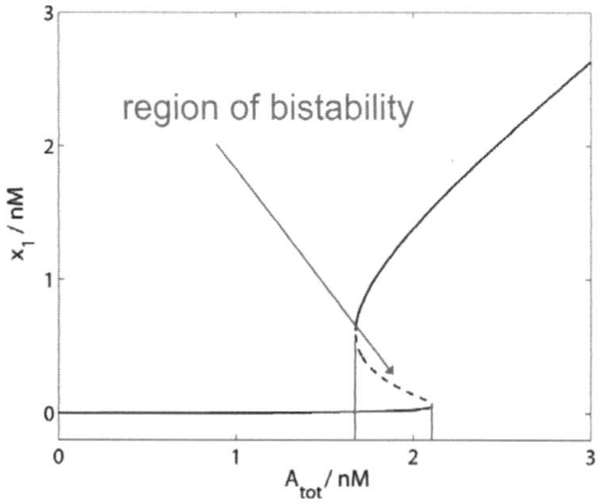

Convex analysis of the bistable network

- Linear stability analysis
  - Jacobi matrix

$$\mathbf{J} = \begin{bmatrix} -\frac{h_1 j_1}{3} & -\frac{h_2 j_1}{3} & 0 & 0 & \frac{h_5 j_1}{3} \\ -\frac{h_1 j_1}{3} & -\frac{h_2 j_1}{3} & \frac{h_3 j_1}{3} & 0 & 1 \\ -\frac{h_1 j_2}{2} & 0 & -h_3(\frac{j_1}{3} + \frac{j_2}{2}) & \frac{h_4 j_2}{2} & \frac{h_5 j_1}{3} \\ \frac{h_1 j_2}{2} & 0 & \frac{h_3 j_2}{2} & -\frac{h_4 j_2}{2} & 0 \\ \frac{h_1 j_1}{3} & -\frac{h_2 j_1}{3} & 0 & 0 & -\frac{h_5 j_1}{3} \end{bmatrix}$$

  - characteristic equation

$$\det \mathbf{J} = \lambda^2(\lambda^3 + \alpha_1 \lambda^2 + \alpha_2 \lambda + \alpha_3) = 0$$

where

$$\alpha_1 = \frac{h_1 j_1}{3} + \frac{h_2 j_1}{3} + \frac{h_3 j_2}{2} + \frac{h_4 j_2}{2} + \frac{h_5 j_1}{3},$$

$$\alpha_2 = \frac{j_1^2}{9}(h_1 h_3 + h_2 h_3 + h_2 h_5 + h_3 h_5) + \frac{j_1 j_2}{6}(h_1 h_3 + h_1 h_4 + h_2 h_3 + h_2 h_4 + h_3 h_4 + h_3 h_5 + h_4 h_5),$$

$$\alpha_3 = \frac{j_1^2 j_2}{18}(h_1 h_3 h_4 - h_1 h_2 h_3 + h_2 h_3 h_4 + h_2 h_3 h_5 + h_2 h_4 h_5 + h_3 h_4 h_5).$$

- The network is linearly unstable if at least one eigenvalue has a positive real part
- Routh-Hurwitz determinants

$$h_1 h_3 h_4 + h_2 h_3 h_4 + h_2 h_3 h_5 + h_2 h_4 h_5 + h_3 h_4 h_5 < h_1 h_2 h_3$$

Convex analysis of the bistable network

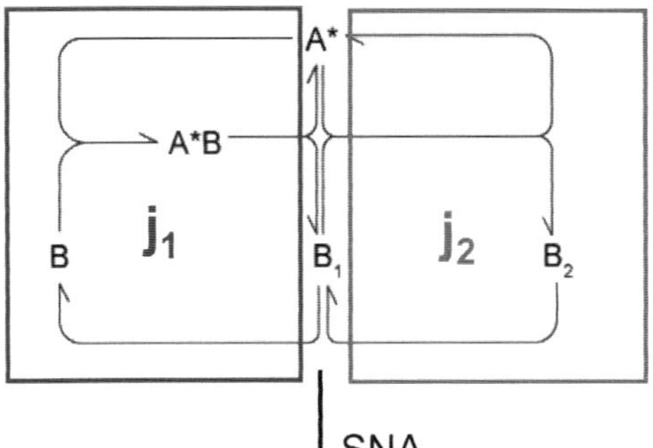

SNA

2 elementary subnetworks

$$\mathbf{E} = [\mathbf{E}_1 \ \mathbf{E}_2] = \begin{bmatrix} \frac{1}{3} & 0 \\ \frac{1}{3} & 0 \\ 0 & \frac{1}{2} \\ 0 & \frac{1}{2} \\ \frac{1}{3} & 0 \end{bmatrix}$$

Weight coefficients of E_1 and E_2

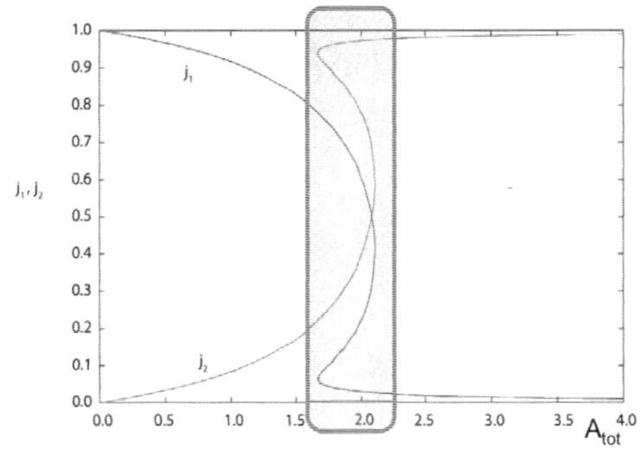

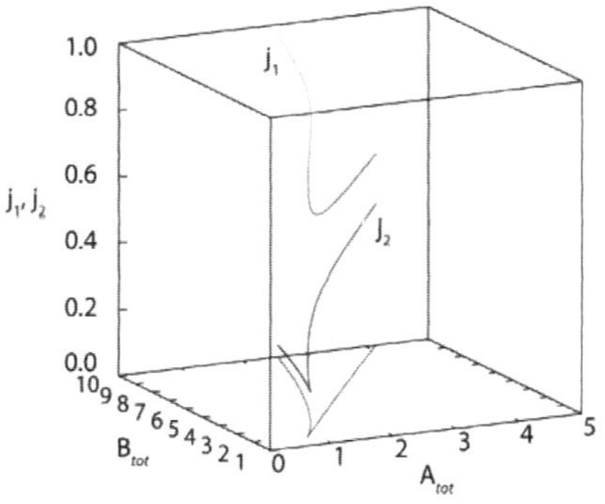

Explanation of the instability via competitive autocatalysis (CA)

Topological fragment of CA (stable)
(I. Schreiber, M. Eiswirth, in preparation)

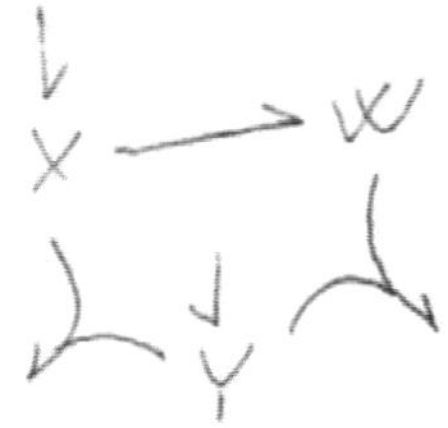

Bistability by substrate inhibition

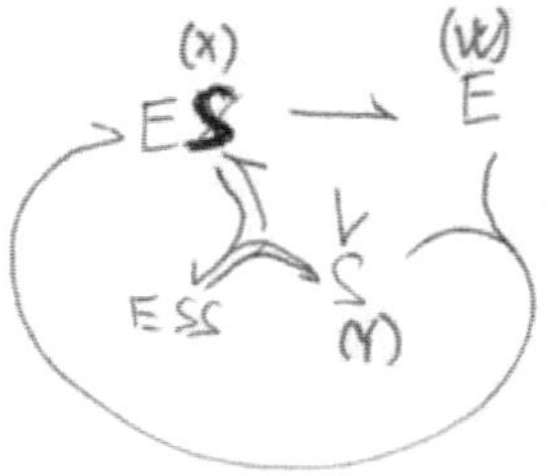

Bistable motif of CA

MAPK reduced network

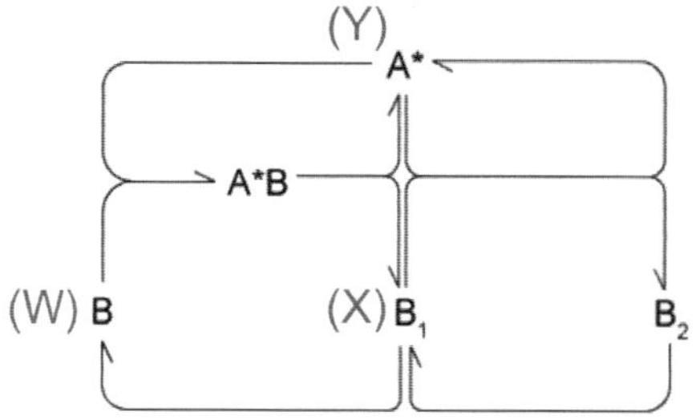

Classification of oscillatory reaction mechanisms

Classification of species: M. Eiswirth, A. Freund, J. Ross, Adv. Chem. Phys. 80, 1991

Essential: must be present for oscillations; if buffered oscillations cease
- type X – autocatalytic species – occur on the cycle
- type Z – species providing negative feedback
- type Y – inhibitory species, take part in the exit reaction
- type W – recovery species, products of tangent or exit reactions

Nonessential: can be buffered with no effect on oscillations
- type a – reactants with a weak feedback
- type b – products with a weak feedback
- type c – intermediates not in the oscillatory pathway

$a \rightleftharpoons Z$
$Z + X \rightarrow 2X$
$X + Y \rightarrow c$
$Y + c \rightarrow$
$Y \rightarrow$
$X \rightleftharpoons b$
$X \rightleftharpoons,\ Y \rightleftharpoons,\ Z \rightleftharpoons$
$a \rightleftharpoons,\ b \rightleftharpoons,\ c \rightleftharpoons$

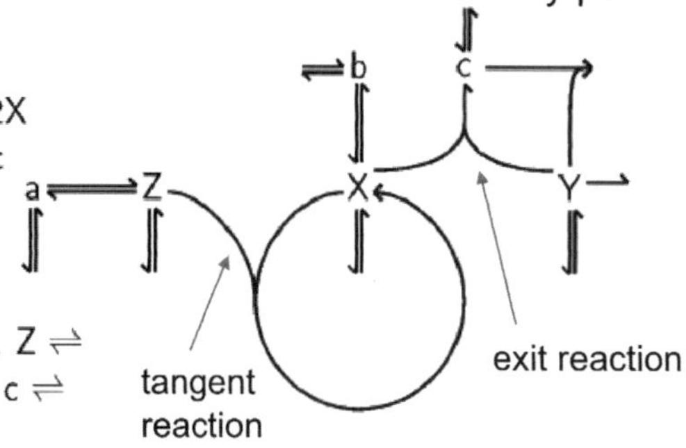

Classification of oscillatory reaction mechanisms

Category 1 – order of autocatalysis = order of exit reaction = 1
Category 2 – order of autocatalysis > 1, no exit reaction

category 1B:

$\square + X \xrightarrow{k_1} 2X + 2Z$
$Z \xrightarrow{k_2} \frac{1}{2}Y$
$Y \xrightarrow{k_3} X$
$X + Y \xrightarrow{k_4}$
$2X \xrightarrow{k_5}$
$X \rightleftharpoons, Y \rightleftharpoons, Z \rightleftharpoons$

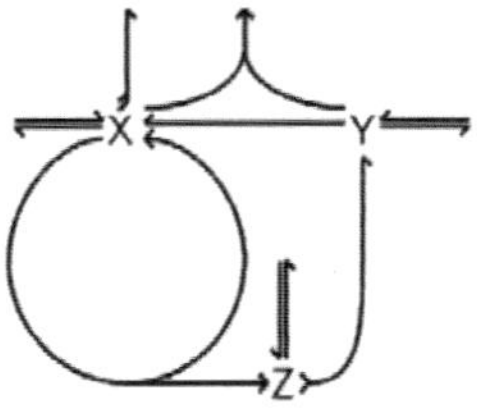

category 1CX:

$Z + X \xrightarrow{k_1} 2X$
$X + Y \xrightarrow{k_2}$
$X \xrightarrow{k_3}$
$Y \xrightarrow{k_4}$
$X \rightleftharpoons, Y \rightleftharpoons, Z \rightleftharpoons$

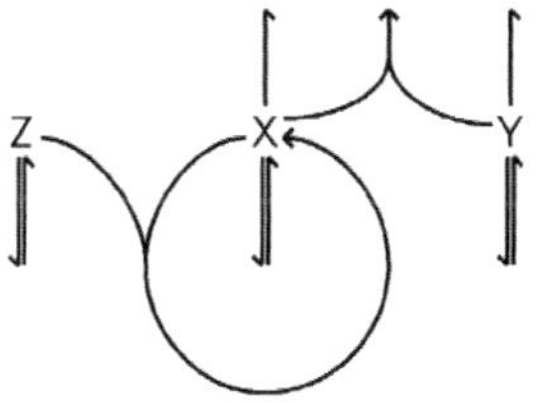

category 2B:

$\square + 2X \xrightarrow{k_1} 3X + Z$
$X + Z \xrightarrow{k_2}$
$Z \xrightarrow{k_3} X$
$2X \xrightarrow{k_4}$
$X \rightleftharpoons$
$Z \rightleftharpoons$

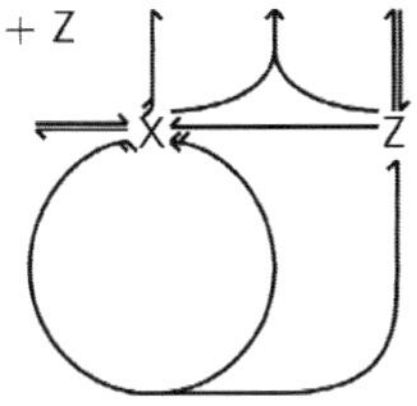

category 2C:

$Z + 2X \xrightarrow{k_1} 3X$
$X \xrightarrow{k_2}$
$X \rightleftharpoons$
$Z \rightleftharpoons$

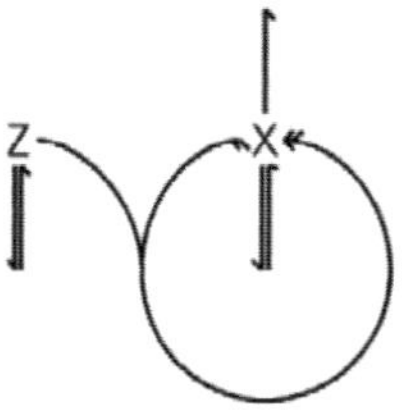

Minimal oscillatory network in the H-F model

X,W – autocatalytic species
Y – exit species
Z – negative feedback species

No autocatalytic cycle !

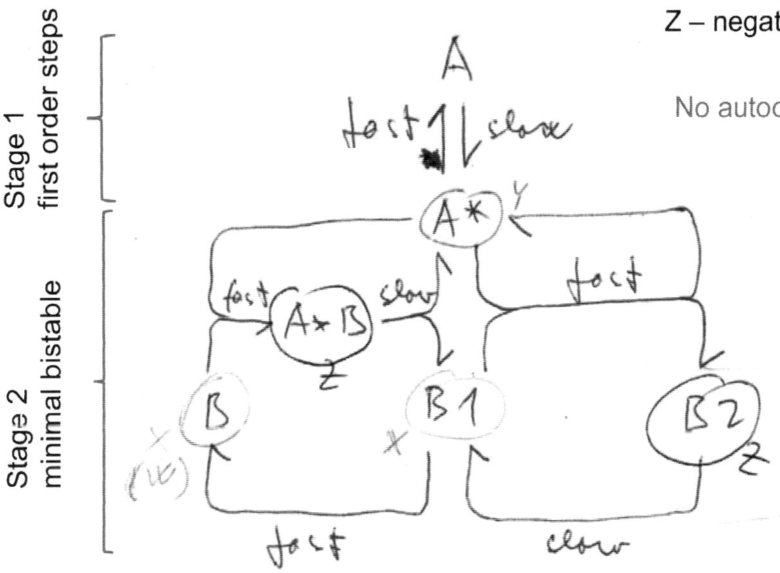

Minimal oscillatory network in the H-F model

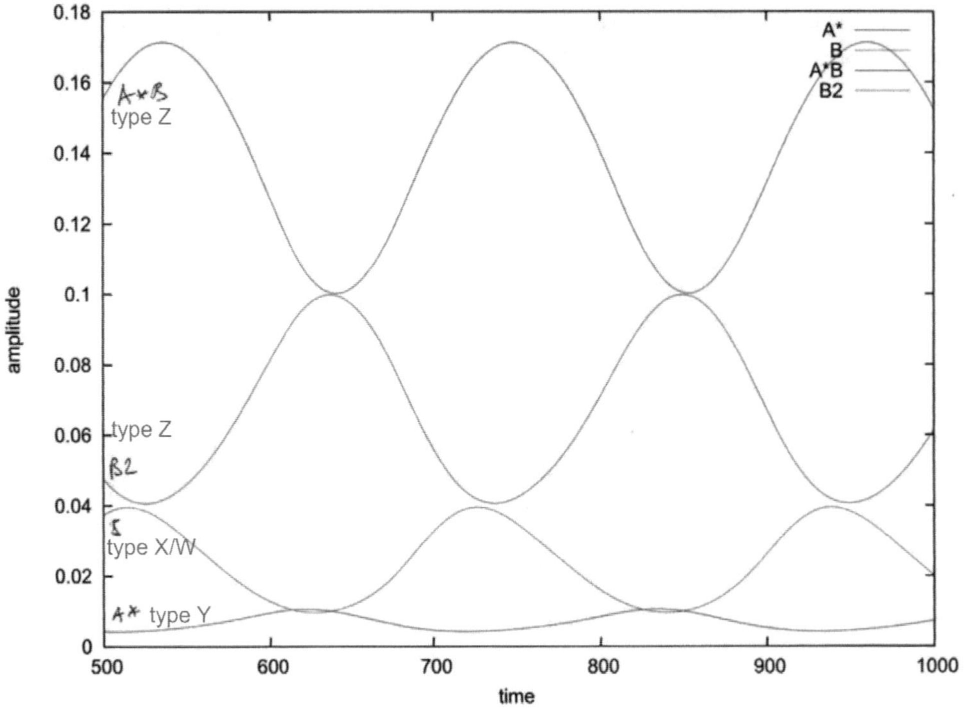

Minimal oscillatory network in the H-F model

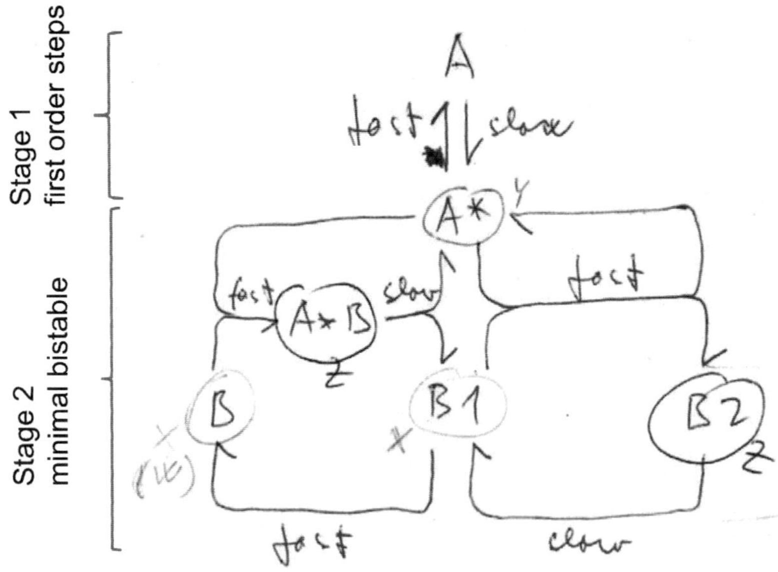

Phase 1 (inhibitory):
initially high B1
is depleted by
reacting with A*
to produce B2;
A*B is also depleted
and helps to drive
production of B2

Phase 2 (activatory):
B1 is regenerated
from B2 while A* is
temporarily low;
B and A*B are likewise
regenerated

Conclusions

- Stoichiometric network analysis of the MAPK reaction mechanism makes it possible to identify positive and negative feedback loops leading to oscillations or bistability.
- Stage 2 is a basis of the core subnetwork for bistability in the Huang-Ferrell model.
- Linear stability analysis provides an inequality for convex parameters j, h which specifies the region of bistability in the kinetic parameter space.
- Extension to the minimal oscillatory network requires additionally Stage 1 of the MAPK cascade, i.e., dynamical activation/deactivation of the triple-kinase form of the MAPK.

The support of the research by the grant of Ministry of Education, Youth and Sport of the Czech Republic KONTAKT II LH14009 is highly acknowledged.

Genome based Metabolic Flux Analysis (MFA) of *Methanococcus maripaludis* for improved methane productivity

Nishu Goyal,[a] Iftekhar A. Karimi,[a*] Zhou Zhi, George[b]

[a]*Department of Chemical and Biomolecular Engineering, 4 Engineering Drive 4, National University of Singapore*

[b]*Department of Civil and Environmental Engineering, 1 Engineering Drive 2, National University of Singapore*

Rising levels of CO_2 in the atmosphere is one of the major problems around the world. At present, CO_2 emissions in atmosphere are minimized by storing or converting the emitted CO_2, which demands compression and is expensive. Recent progress shows that microbial conversion of CO_2 to methane is a novel biocatalytic approach for carbon capture and conversion to fuel that seems relatively economical.

Hydrogenotrophic methanogens are archaebacteria that utilize CO_2 and convert it to methane for ATP synthesis. *Methanococcus maripaludis,* a hydrogenotrophic methanogen, appears to be a highly promising methanogen for large-scale utilization in industries due to its inherent ability to utilize CO_2 and N_2, which are major components of flue gas. In addition, it's short doubling time of 2h under mesophilic conditions and availability of tools for genetic manipulations offer additional advantages. Even though the microbe has been thoroughly studied, the studies on improved methane productivity are lacking in the literature.

In order to study genetic and environmental parameters affecting methanogenesis, this work present a first constraint-based genome-scale metabolic model of *M. maripaludis* S2. This model consists of 570 reactions, 556 distinct metabolites, and 518 genes across 52 pathways and covers 30% of open reading frames (ORFs). The draft model was constructed with the help of online databases and further refined using literature and annotated or proteomic evidences. The model was successfully validated using biomass growth data and phenotypic studies present in the literature. Using the experimentally measured CO_2 uptake rate as input, the model predicts specific growth rate, methane evolution rate, hydrogen uptake rate in addition to intracellular fluxes across all metabolic pathways under different media conditions. Based on the model findings, we were able to propose a hypothesis that different carbon and nitrogen sources influence growth and methanogenesis significantly. The comparison of fluxes with *Methanosarcina barkeri*, *Methanosarcina acetivorans*, and *Sulfolobus solfataricus* P2 showed *M.*

maripaludis S2 to be a better organism for producing methane from CO_2. *In silico* flux analysis of metabolic network using various metabolic engineering approaches such as gene(s) knockouts, over expression of certain gene(s), etc. helped us in identifying potential genes targets for improved methane productivity. This model also provides a framework for the comprehensive study of existing or new metabolic processes, effects of media components, effect of inhibitory metabolites etc. in different strains of *M. maripaludis* and in other hydrogenotrophic methanogens. Experimental studies are underway to validate our model predictions.

Keywords: Biocatalysis, Flue gas, Hydrogenotrophic methanogens, *M. maripaludis*, Genome-scale model, Flux balance analysis

Impacts of Low Peclet Number Flow and Oscillatory Flow on Autocrine Signaling

Michal Přibyl & Igor Schreiber

Department of Chemical Engineering
Institute of Chemical Technology, Prague

pribylm@vscht.cz

2014 AIChE Annual Meeting, Atlanta, Nov 16-21, 2014

Outline

- Convection in cellular systems
- EGFR mediated cellular signaling
- Mathematical model
- Effects of steady convection on signal propagation
- Effects of oscillatory convection on signal propagation
- Conclusions

Department of Chemical Engineering, ICT Prague

Convection in cellular systems

Peclet Number

$$\mathrm{Pe} = \frac{uL}{D} = \frac{u}{\frac{D}{L}} = \frac{\text{convection intensity}}{\text{diffusion intensity}}$$

- Dominant in large vessels
  - Pe » 1
  - Vascular and lymphatic system
  - Long-range endocrine communication

- Important in small capillaries and compact tissues?
  - Characteristic dimension $L \sim 1\ \mu\mathrm{m}$
  - Ligand diffusivity $D \sim 1\times10^{-11}\ \mathrm{m^2 s^{-1}}$
  - Velocity $u \sim 10\ \mu\mathrm{m\ s^{-1}}$
  - Short-range communication (autocrine, paracrine, juxtacrine)
  - Flow intensity comparable with diffusion transport intensity, Pe ~ 1

Department of Chemical Engineering, ICT Prague

Convection transport – role in tissues

Capillary morphogenesis

Semino et al., Experimental Cell Research 312, 2006, 289– 298

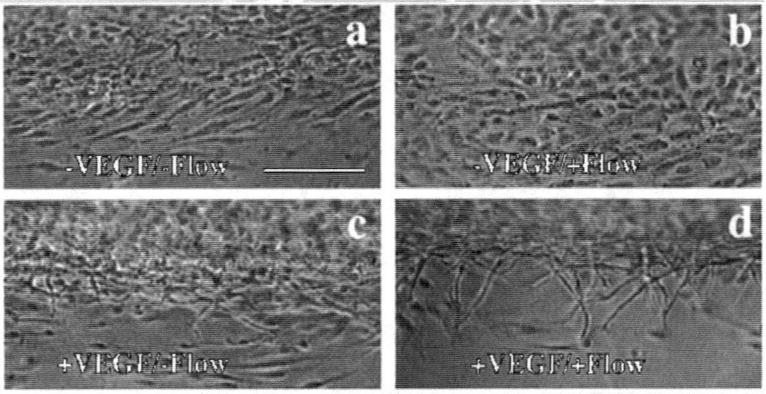

Helm et al. PNAS 102, 2005, 15779–15784

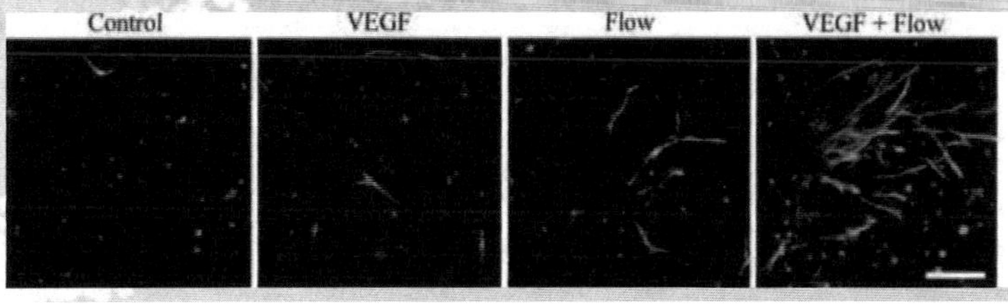

Department of Chemical Engineering, ICT Prague

Convective transport – role in tissues

- Morphogenesis of vascular and lymphatic systems
- Cell (lymphocytes, cancer cells) migration in lymphatic capillaries – convection deforms gradients of chemotactic proteins
- Convection induces stress in tissue – activation of integrins (cellular receptors)
- Formation of left-right axis in embryos
- Nutrient supply in cartilages and bones
- Oscillatory flow – induced by pressure variations due to heart action, walking, breathing …
- Specific responses of cellular cultures to different flow conditions in the extracellular space (expression of atherosclerosis protective genes under steady laminar flow but not under oscillatory flow)

Department of Chemical Engineering, ICT Prague

Short-Range Cellular Signaling

- Phases
  - Ligand release
  - Ligand transport
  - Ligand-receptor binding
  - Intracellular processing
- Effects
  - Proliferation
  - Differentiation
  - Apoptosis
  - Movement
  - Stable regime

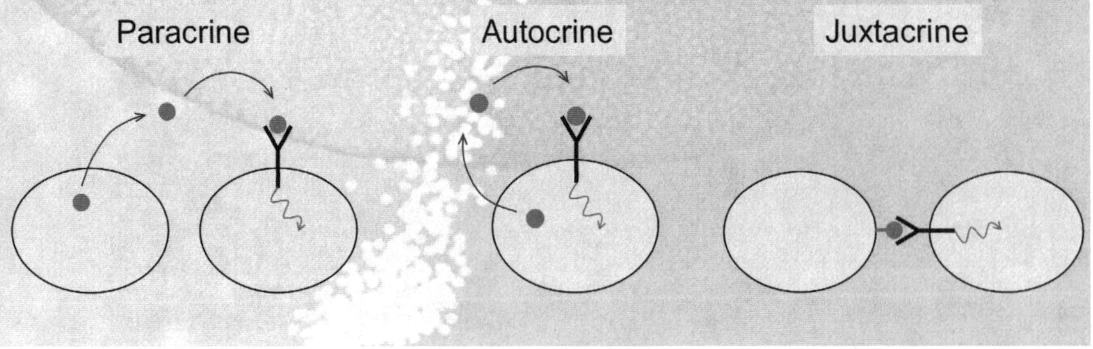

Department of Chemical Engineering, ICT Prague

Epidermal growth factor receptor signaling

- Transmembrane tyrosine kinase receptors
- Extracellular domain
  - Ligand binding
  - Ligand transport
    - Diffusion
    - Convection
  - Complex dimerization
- Intracellular domain
  - Tyrosine kinase autophosphorylation
  - Ras, Raf proteins
  - MAPK cascade
  - Targeting

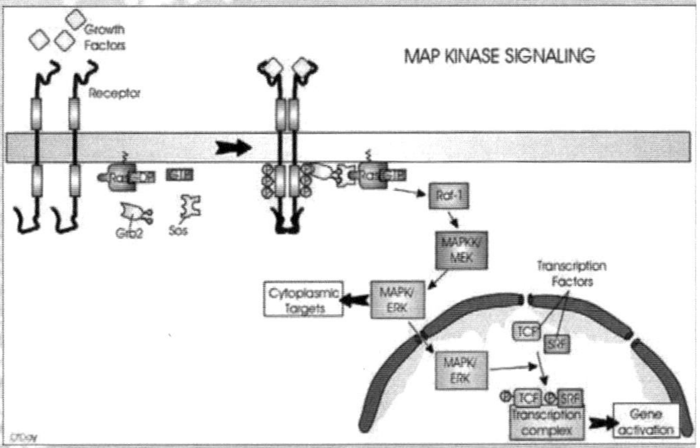

www.utm.utoronto.ca/~w3bio315/picts/lectures/lecture21/Mapkst1.jpg

Department of Chemical Engineering, ICT Prague

Epithelia

One-layer epithelium

Gut epithelium

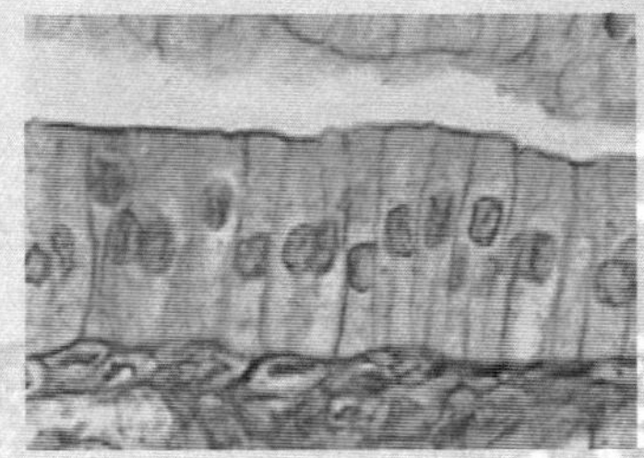

Coleman, Technion

Multi-layer epithelium

Tongue epithelium

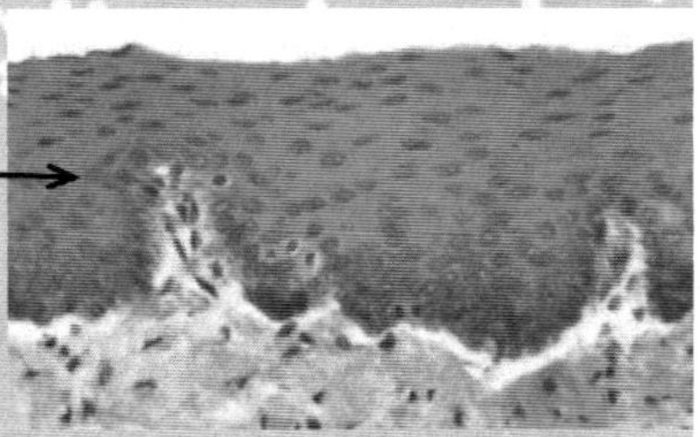

Coleman, Technion

One-layer epithelium

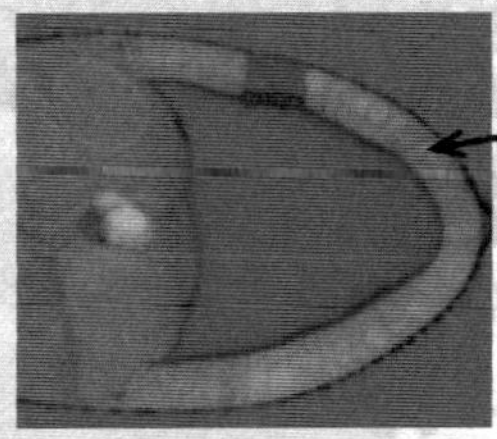

Follicular epithelium
Drosophila melanogaster

Baum, *Nature Cell Biol.* 3, 2001, 883.

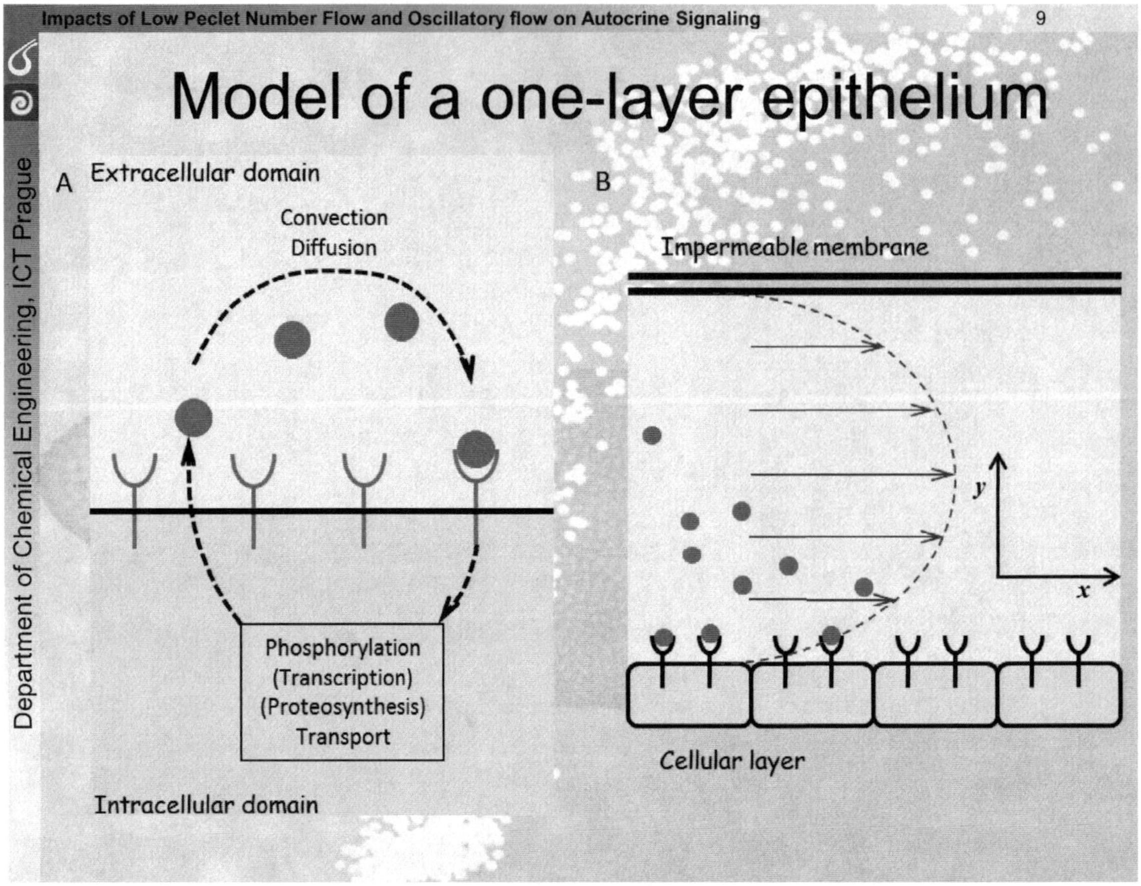
Impacts of Low Peclet Number Flow and Oscillatory flow on Autocrine Signaling
9
Department of Chemical Engineering, ICT Prague
Model of a one-layer epithelium
A
Extracellular domain
Convection
Diffusion
Phosphorylation
(Transcription)
(Proteosynthesis)
Transport
Intracellular domain
B
Impermeable membrane
y
x
Cellular layer

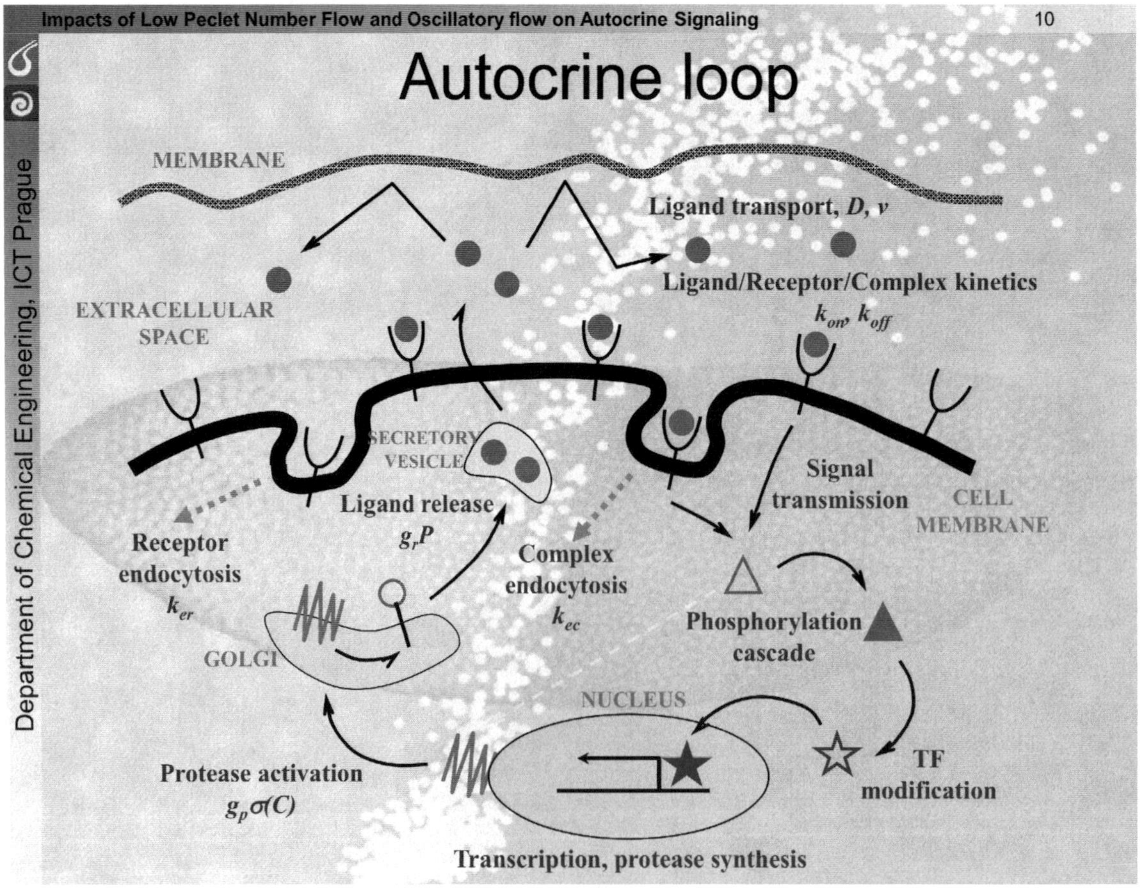
Impacts of Low Peclet Number Flow and Oscillatory flow on Autocrine Signaling
10
Department of Chemical Engineering, ICT Prague
Autocrine loop
MEMBRANE
Ligand transport, D, v
Ligand/Receptor/Complex kinetics
k_{on}, k_{off}
EXTRACELLULAR SPACE
SECRETORY VESICLE
Ligand release
g_rP
Signal transmission
CELL MEMBRANE
Receptor endocytosis
k_{er}
Complex endocytosis
k_{ec}
Phosphorylation cascade
GOLGI
NUCLEUS
TF modification
Protease activation
$g_p\sigma(C)$
Transcription, protease synthesis

Department of Chemical Engineering, ICT Prague

Mathematical model

Ligand balance in extracellular space together with top and bottom boundary conditions

$$\frac{\partial S}{\partial t} = -v_x(y,t)\frac{\partial S}{\partial X} + D\frac{\partial^2 S}{\partial X^2} + D\frac{\partial^2 S}{\partial Y^2}$$

$$Y = H \qquad \frac{\partial S}{\partial Y} = 0,$$

$$Y = 0 \qquad -D\frac{\partial S}{\partial Y} = k_{off}C - k_{on}SR + g_r P,$$

Balances of receptor concentration and ligand-receptor complex concentration

$$\frac{\partial R}{\partial t} = Q_r - k_{er}R + k_{off}C - k_{on}SR.$$

$$\frac{\partial C}{\partial t} = k_{on}SR - \left(k_{off} + k_{ec}\right)C,$$

Activity of ligand releasing protease

$$\frac{\partial P}{\partial t} = -k_p P + g_p \sigma(C - C_T),$$

Pribyl et al. Biophysical Journal 84, (2003) 883–896.

Department of Chemical Engineering, ICT Prague

Dimesionless model

Ligand

$$\tau_S\frac{\partial \tilde{S}}{\partial \tilde{t}} + \tilde{v}_x(\tilde{y},\tilde{t})\frac{\partial \tilde{S}}{\partial \tilde{x}} = \frac{\partial^2 \tilde{S}}{\partial \tilde{x}^2} + \frac{1}{\alpha^2}\frac{\partial^2 \tilde{S}}{\partial \tilde{y}^2}$$

$$-\frac{\partial \tilde{S}}{\partial \tilde{y}} = \alpha\left[(1-\beta_S)\tilde{C} - \tilde{S}\tilde{R} + \beta_S\tilde{P}\right], \quad \tilde{y} = 0$$

$$\frac{\partial \tilde{S}}{\partial \tilde{y}} = 0, \quad \tilde{y} = 1$$

Receptor

$$\tau_R\frac{\partial \tilde{R}}{\partial \tilde{t}} = 1 + \gamma\left[-\tilde{S}\tilde{R} + (1-\beta_S)\tilde{C}\right] - \tilde{R}$$

$$\tau_C\frac{\partial \tilde{C}}{\partial \tilde{t}} = \tilde{S}\tilde{R} - \tilde{C}$$ Complex

$$\frac{\partial \tilde{P}}{\partial \tilde{t}} = \tilde{\sigma}\left(\tilde{C} - \tilde{C}_T\right) - \tilde{P}$$ Protease

Dimesionless parameters

- α - Damköhler number
- β_S - Relative strength of endocytosis
- γ - Ligand to receptor ratio of generation rates
- τ_S - Ligand time scale
- τ_C - Complex time scale
- τ_R - Receptor time scale
- $\tilde{C}_T$ - Protease activation threshold

Department of Chemical Engineering, ICT Prague

Two interpretations of α

- Geometric ratio
  - h – dimension of the vertical gap above the cellular layer
  - L_x – characteristic spatial range of ligand trafficking above epithelium

$$\alpha = \frac{h}{L_x}$$

- Damköhler number
  - $k_{on}R_0$ – rate of ligand-receptor binding
  - D/h – rate of ligand diffusion

$$\alpha = \frac{k_{on}R_0}{\dfrac{D}{h}}$$

Department of Chemical Engineering, ICT Prague

Dimesionless convection velocity – function of the Peclet number

- Plug flow

$$\tilde{v}_x(\tilde{y}) = \mathrm{Pe}$$

- Laminar flow - parabolic

$$\tilde{v}_x(\tilde{y}) = 6\mathrm{Pe}(\tilde{y} - \tilde{y}^2)$$

- Oscillatory flow - harmonic

$$\tilde{v}_x(\tilde{t}) = \mathrm{Pe}(\tilde{t}) = \mathrm{Pe}_A \sin(2\pi \tilde{f}\tilde{t})$$

Taylor dispersion

$$D_T = D_L(1 + \alpha^2\mathrm{Pe}^2/210)$$

For Pe ~ 1 and α ~ 1 , the Taylor dispersion effect is negligible and laminar flow is well approximated by plug flow.

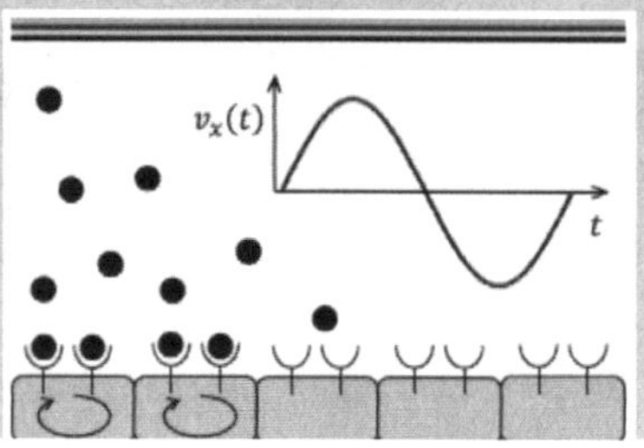

Department of Chemical Engineering, ICT Prague

Autocrine communication is mediated by traveling concentration waves

$$\xi \equiv \tilde{x} - \tilde{u}\tilde{t}$$

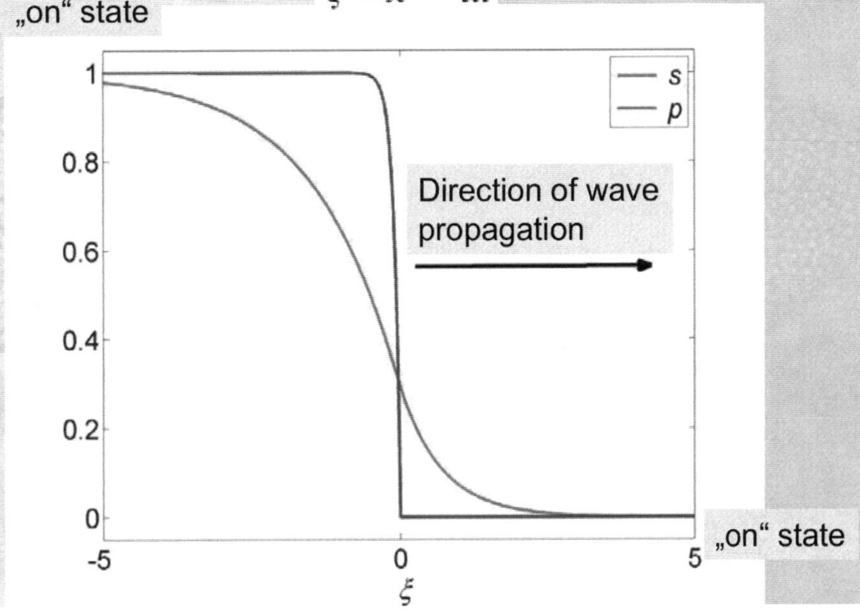

Department of Chemical Engineering, ICT Prague

Robust property of autocrinne communication

Dependence of velocity propagation on the relative binding strength α exhibits single maximum

Binding is too week to start intracellular cascades $\alpha \to 0$
Binding is too intensive to allow for ligand trafficking $\alpha \to \infty$

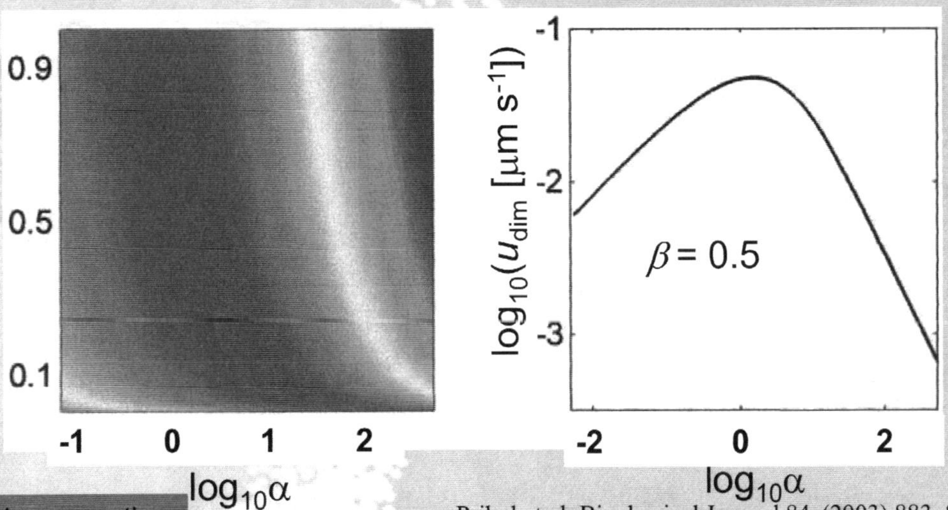

No convection

Pribyl et al. Biophysical Journal 84, (2003) 883–896.

Department of Chemical Engineering, ICT Prague

Effect of low Peclet number flow on the velocity of signal propagation

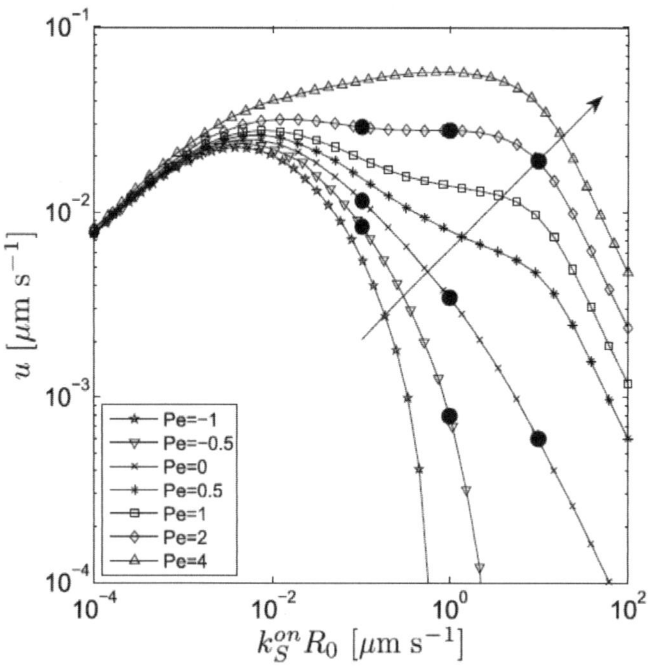

- Region of high velocities becomes much broader for $\mathrm{Pe} > 0$
- Signaling is faster by more than one order of magnitude $\mathrm{Pe} > 0$
- Propagation is slowed down, stopped or reversed $\mathrm{Pe} < 0$

Nebyla et al. Biophysical Journal 102, (2012) 990–1000.

Department of Chemical Engineering, ICT Prague

Effect of low Peclet number flow on the direction of signal propagation

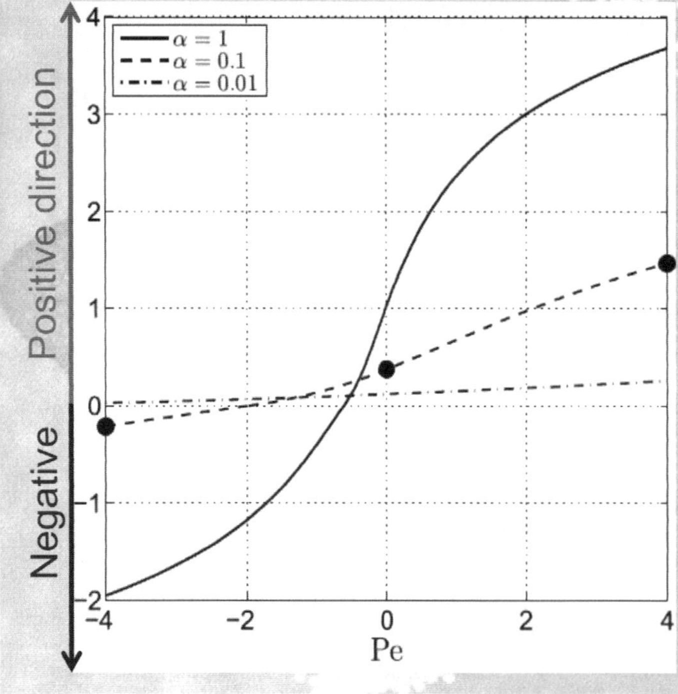

- Convective transport significantly affects only such epithelia, in which the characteristic dimension of extracellular gap is larger than the spatial extent of ligand diffusion trafficking
- $\alpha > 1$

Nebyla et al. Biophysical Journal 102, (2012) 990–1000.

Department of Chemical Engineering, ICT Prague

Effect of low Peclet number flow – character of propagating chemical waves

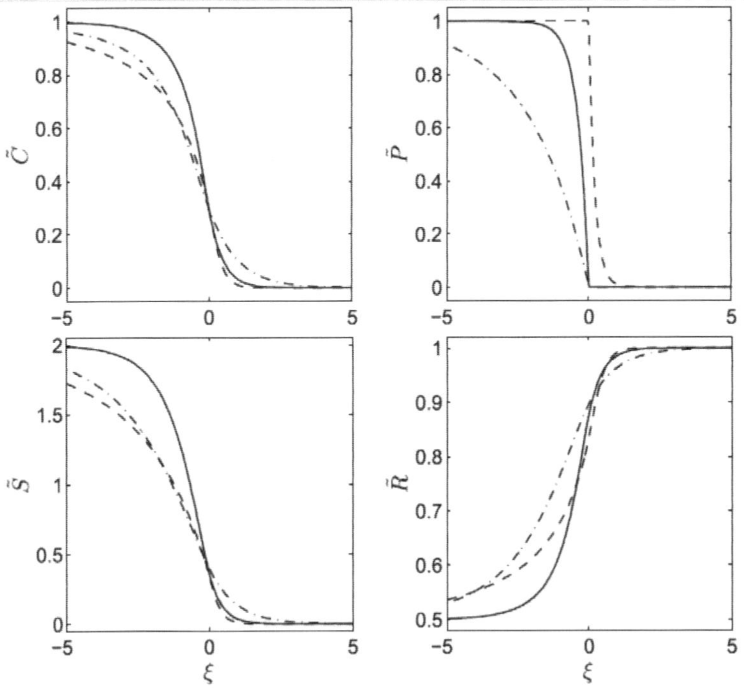

- Solid line
  $\mathrm{Pe} = 0, \tilde{u} = 0.38$
- Dash-dotted line
  $\mathrm{Pe} = +4, \tilde{u} = 2.06$
- Dashed line
  $\mathrm{Pe} = -4, \tilde{u} = -0.22$

Department of Chemical Engineering, ICT Prague

Oscillatory convective flow – character of propagating chemical waves

Propagation velocity is a periodic function of time because $\mathrm{Pe}(\tilde{t}) = \mathrm{Pe}_A \sin(2\pi\tilde{f}\tilde{t})$

During one period – concentration wave changes its shape

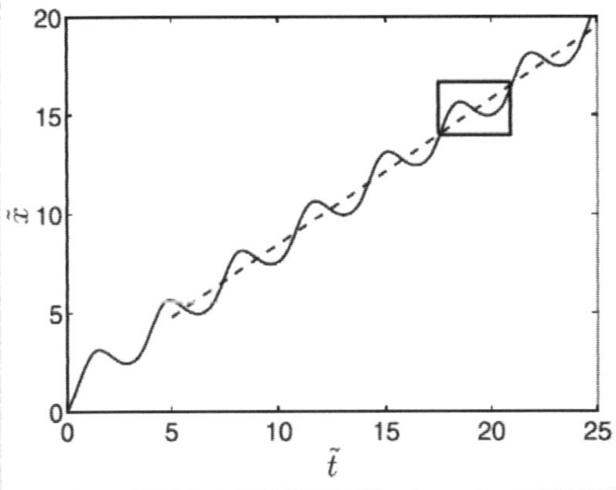

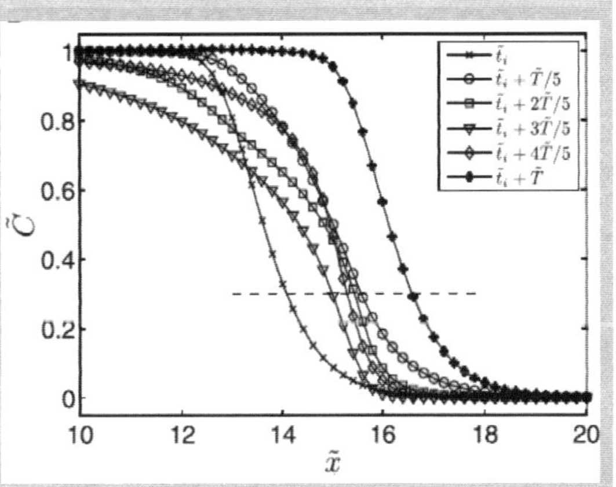

Nebyla et al. Biophysical Journal 105, (2013) 818–828.

Oscillatory convective flow – frequency characteristics

- There two asymptotes on the frequency characteristics
- Lower asymptote (short period, high frequency)
- Upper asymptote (long period, low frequency)
- Nonlinear velocity dependencies between asymptotes
- A, B, C – different sets of kinetic and transport parameters, $\mathrm{Pe}_A = 4$

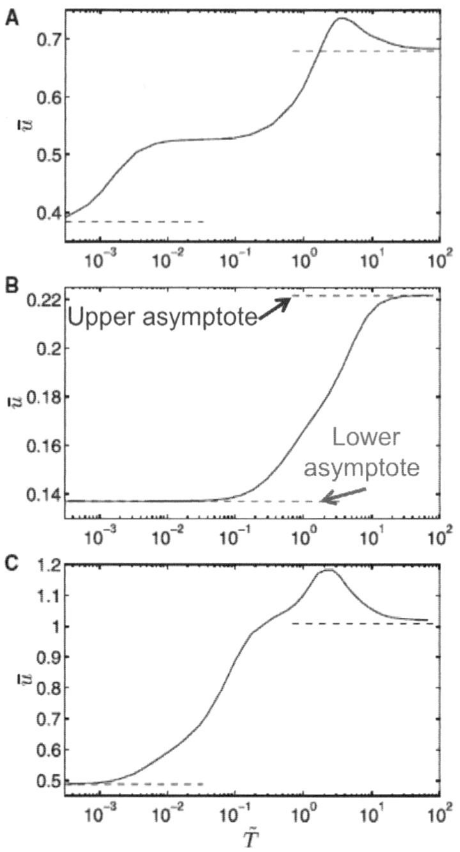

Department of Chemical Engineering, ICT Prague

Conclusions

- Low Peclet number convection in compact tissues can significantly accelerate the propagation of chemical signals if oriented downstream
- Orientation against the propagating signal can lead to deceleration, stopping or reversal in signaling
- Convection can become significant only if the characteristic dimension of extracellular space is larger that the spatial extent of ligand trafficking
- Harmonic oscillatory convection accelerates autocrine signaling
- The acceleration is limited by an asymptotic value of the propagation velocity
- The asymptotic value results from the convex shape of the velocity versus Peclet number dependence in non-oscillatory regimes

The support of the research by the grants of Ministry of Education, Youth and Sport of the Czech Republic KONTAKT II LH14009 is highly acknowledged.

392983 Constructing Influenza induced Host Response Models from Microarray Data

Abstract:

The influenza virus remains a potent threat to public health. Recently, researchers have considered functions of the host immune response as potential therapeutic targets but limited data, strain specific responses, and high variation of the host response between typical cell and animal infection models has limited the development of host based drug therapies. In our group, we have focused on developing a technique referred to as systems inference microarray analysis (SIMA) for developing prediction capable, species specific models of the immune response using whole genome data. We exploit biological mechanisms such as gene co expression to reduce whole genome data down to clusters of genes which represent the activity of biological pathways or processes. A collection of nonlinear equations which are strictly a function of virus growth are then fit to the expression dynamics and used to predict the host gene response in new infections. Applying this technique to time course microarray data developed from mouse lung infected with three different influenza isolates, we have identified novel mechanisms regulating a key inflammatory event – the cytokine storm. Further, applying the approach to cell plate data, we have found that the dynamics of inflammatory and immune response pathways are highly variant. This suggests that *in vitro* data is a poor biological model of the *in vivo* inflammation response. We believe applying the SIMA approach to a more extensive data set will lead to a more complete understanding of infection induced inflammation and guide the development of future treatments.

The Differential Roles of GABA and VIP in Synchronization and Entrainment of the Suprachiasmatic Nucleus: A Mathematical Modeling Study

Nathaniel J. Kingsbury[1] Stephanie Taylor[2] and Michael A. Henson[1]
[1]Department of Chemical Engineering, University of Massachusetts, Amherst, MA 01007
[2]Department of Computer Science, Colby College, Waterville, ME 04901

In mammals, the suprachiasmatic nucleus (SCN) of the hypothalamus is the dominant circadian pacemaker that drives daily rhythms in behavior and physiology. The SCN consists of approximately 20,000 neurons that can act as individual circadian pacemakers but instead are coupled together to produce a robust and synchronized overall rhythm. Two key neurotransmitters involved in the generation and maintenance of SCN rhythms are γ-aminobutyric acid (GABA) and vasoactive intestinal polypeptide (VIP). While numerous studies have shown that VIP is essential for intercellular synchronization, the role of GABA in the SCN remains controversial. Various experimental studies have shown that GABA can be inhibitory or excitatory, can be primarily active at night or the day, and can promote synchronization or not promote synchronization. These seemingly inconsistent results motivate the need for new experimental and mathematical tools that allow a more complete and careful analysis of GABA signaling in the SCN. Because GABA is the major neurotransmitter in animal brains, improved understanding of GABA circadian signaling could have wide ranging implications for other brain regions.

The goal of this study was to use mathematical modeling to better understand the differential roles of GABA and VIP signaling in the SCN. The research was designed to test our hypothesis that GABA improves the ability of cells communicating through VIP to phase delay or phase advance their target cells. Computational analysis was performed with our recently developed multicellular model that integrates GABA, VIP and photic signaling across coupled populations of heterogeneous SCN neurons (Vasalou et al., 2011). Simulations with assorted network topologies were conducted to investigate the effects of eliminating GABA or VIP signaling on synchronization and period adjustment in constant darkness and entrainment to different light-dark schedules. Supported by NIH Grants GM078993 and 1R01GM096873-01.

Vasalou C, Herzog ED, Henson MA. Multicellular Model for Intercellular Synchronization in Circadian Neural Networks. Biophysical Journal, 101(1): 12-20, 2011.

Development of amplitude response curves for single-cell and population-level circadian systems

Peter C. St. John and Francis J. Doyle III

Department of Chemical Engineering, University of California, Santa Barbara

Circadian rhythms are oscillations in biological systems that help organisms anticipate changes in the day/night cycle. These oscillations are typically coordinated by intricate genetic regulatory networks that generate tunable oscillations through time-delayed negative feedback loops. In addition to serving as an excellent example of a functional genetic circuit [1], circadian rhythms form an integral part of the mammalian metabolic regulatory network. Consequently, disruptions to the circadian clock often manifest in metabolic disease [2]. Since circadian rhythms are damped by factors such as age [3], diet [4], and shift work, an understanding of the forces that regulate clock amplitudes is essential. However, such an understanding is obstructed by the multi-scale nature of the circadian system. In mammals, locomotor rhythms are governed by the suprachiasmatic nucleus (SCN), which serves as the body's master pacemaker. Metabolic rhythms and other physiological outputs of the circadian clock are mediated by oscillating cells in peripheral tissues. Unlike the SCN, cells in tissues such as the liver and pancreas lack the coupling mechanisms necessary for spontaneous synchronization. As a result, cultured peripheral tissues display rhythms which damp appreciably in only a few days, as stochastic effects cause an initially entrained population of cells to desynchronize. The amplitude of circadian rhythms in metabolic tissues is therefore determined both by the genetic oscillator within each cell and the synchrony of the overall population.

Due to its inherent complexity, circadian gene expression data is often understood with the aid of kinetic models. Such models, which typically take the form of coupled ordinary equations (ODEs), have helped explain how perturbations to individual genetic components effect changes in systems-level properties such as phase or period [5]. Definitions and computational techniques form the foundation of this analysis, allowing efficient parameter estimation and direct comparison between models and experiments. However, these techniques have tended to focus on changes in phase or period, for which data has been historically been available and which are more consistent between the population and single-cell level. For instance, the phase response curve (PRC) has been used to characterize the entrainment behavior of both experimental and mathematical systems [6]. Definitions of the PRC for mathematical systems have therefore been developed [7, 8], paving the way for a more thorough understanding of phase entrainment and allowing the creation of control strategies to achieve therapeutic effects [9]. Similarly, the period sensitivity, which measures how the oscillatory period changes with respect to a permanent change, has been used to characterize the effects gene silencing and pharmacological changes. However, metrics for calculating amplitude change are lacking.

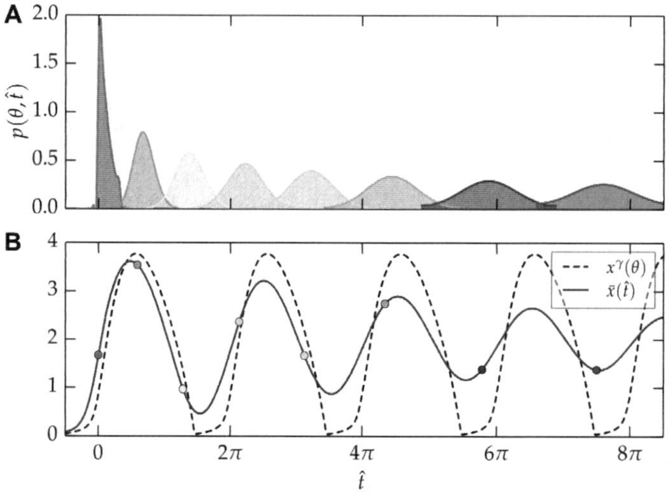

Figure 1: Modeling populations of non-interacting oscillators with the use of continuous probability distributions. (A) A population of oscillators can be described using a distribution to describe the probability of finding an oscillator with a particular phase, $p(\theta, \hat{t})$. As time increases, the probability density follows a convection-diffusion equation, in which the mean phase advances while stochastic effects drive the population to desynchronize. (B) As a result of the desynchronization, population-level rhythms, $\bar{x}(t)$, gradually loose amplitude even though the single-cell rhythm amplitude $x^{\gamma}(t)$ remains constant.

In this study, we provide new tools and definitions to quantify amplitude change in a population of oscillators, and determine what factors dominate amplitude change under certain conditions. Specifically, we use ODE limit-cycle models ${}^{dx}/_{d\hat{t}} = f(x, p)$, which describe the behavior of each individual cell, coupled with a continuous probability density function $p(\theta, \hat{t})$ to describe the synchrony of the population (Figure 1). The population synchrony gradually fades with time, as biological noise causes the phases of each individual cell to diffuse across $(0, 2\pi]$ according to:

$$\frac{\partial p}{\partial \hat{t}} = \frac{\partial p}{\partial \theta} + d\frac{\partial^2 p}{\partial \theta^2} \tag{1}$$

Following a perturbation with a finite duration, such as those encountered in entraining to an external signal, a limit cycle oscillator $x^{\gamma}(\theta)$ undergoes a transient change in amplitude as it returns to the limit cycle with a phase shift $\Delta\theta$. The size of this amplitude deviation, just as the direction of the phase shift, depends on the initial phase at which the perturbation is applied. We propose a variance-based amplitude metric (Eq. 2) to quantify the deviation of the perturbed trajectory $x(t)$ from a reference trajectory $y(t)$, and use this metric to define an amplitude response curve (ARC), analogous to the PRC.

$$\Delta A(x(t), y(t)) := \int_0^\infty (x(t) - \mu)^2 - (y(t) - \mu)^2 \, dt \tag{2}$$

We also derive sensitivity-analysis based methods to efficiently calculate the differential ARC, in which $S(\hat{t})$ is the ODE sensitivity matrix:

$$\frac{dA}{dx} := \lim_{\Delta x(0)\to 0} \frac{\Delta A\left(x(\hat{t}),\ x^{\gamma}(\hat{t}+\theta_0+\Delta\theta)\right)}{\Delta x(0)} \tag{3}$$

$$\frac{dA}{dx} = \int_0^{\infty} 2\left(S(\hat{t}) - \hat{f}(x^{\gamma}(t_\theta))\frac{d\theta}{dx}\right)\left(x^{\gamma}(t_\theta) - \mu\right)\, dt \tag{4}$$

When the mean expression level of a population of oscillators is considered, an additional change in amplitude is incurred from the change in population synchrony which persists until the synchrony is changed by subsequent perturbations. These changes, which are controlled by the phase transition curve $g(\theta)$, can be efficiently approximated by averaging the effect on the mean phase and resultant length over the phase probability density function:

$$z := \int_0^{2\pi} e^{i\theta} p(\theta,\hat{t})\, d\theta \tag{5}$$

$$\hat{z} := \int_0^{2\pi} e^{i\theta} \hat{p}(\theta,\hat{t})\, d\theta = \int_0^{2\pi} e^{ig(\theta)} p(\theta,\hat{t})\, d\theta \tag{6}$$

The amplitude and phase change of the population can then be found by comparing the vector before and after perturbation:

$$\Delta\bar{\theta} = \angle z - \angle\hat{z} \tag{7}$$

$$\Delta\rho = |z| - |\hat{z}| \tag{8}$$

There is therefore a separation in timescales between the effects on clock amplitude mediated at the single-cell level and those mediated by population synchrony. By developing appropriate definitions for amplitude, control strategies might be coupled to mathematical models to develop pharmacological or behavioral therapies which maximize amplitude.

References

[1] Doyle F. J., Gunawan R., Bagheri N., Mirsky H. P., To T.-L. Circadian rhythm: A natural, robust, multi-scale control system. *Comput. Chem. Eng.* 2006;30:1700–1711.

[2] Marcheva B., Ramsey K. M., Buhr E. D., et al. Disruption of the clock components clock and bmal1 leads to hypoinsulinaemia and diabetes. *Nature.* 2010;466:627–631.

[3] Chang H.-C., Guarente L. Sirt1 mediates central circadian control in the scn by a mechanism that decays with aging. *Cell.* 2013;153:1448–1460.

[4] Hatori M., Vollmers C., Zarrinpar A., et al. Time-restricted feeding without reducing caloric intake prevents metabolic diseases in mice fed a high-fat diet. *Cell Metab.* 2012;15:848–860.

[5] St John P. C., Hirota T., Kay S. A., Doyle F. J. Spatiotemporal separation of per and cry posttranslational regulation in the mammalian circadian clock. *Proc. Natl. Acad. Sci. U. S. A.* 2014;111:2040 2045.

[6] Taylor S. R., Doyle F. J., Petzold L. R. Oscillator model reduction preserving the phase response: application to the circadian clock. *Biophys. J.* 2008;95:1658–1673.

[7] Kramer M., Rabitz H., Calo J. Sensitivity analysis of oscillatory systems. *Appl. Math. Model.* 1984;8:328–350.

[8] Gunawan R., Doyle F. J. Isochron-based phase response analysis of circadian rhythms. *Biophys. J.* 2006;91:2131–2141.

[9] Bagheri N., Stelling J., Doyle F. J. Circadian phase resetting via single and multiple control targets. *PLoS Comput. Biol.* 2008;4:e1000104.

389809 Inputs Drive Variability of Regulatory Network Phenotypes

Abstract:

Current high-throughput technologies have allowed us to examine single cell behavior extensively at the molecular level. The ability to develop such comprehensive data sets provides the opportunity to explore and identify system-wide gene expression behavior associated with individual cellular states. Accordingly, the ability to generate causal input-output gene regulatory relationships will provide significant insight into how these molecular mechanisms drive cellular state and function. We aim to develop a network inference technique that models causal input-output gene relationships from single cell transcriptomic data sets. Originally, analyzing such data sets was complicated by the extensive transcriptional variability observed across cells within a particular phenotype. However, correlated gene expression observed in heterogeneous transcriptional behavior across single cells provides clues to underlying regulatory network structures. Moreover, single cell transcriptomic data sets provide a rich experimental sampling of transcriptional profiles that can be used to infer regulatory networks that drive these correlated relationships.

We obtained a high-throughput *in vivo* transcriptomic measures taken from single brainstem neurons of rats undergoing a physiologically induced hypertensive challenge. Our analysis showed that correlated gene expression patterns underlie a distribution of neuronal subtypes[1]. In the present study we examined this single-neuron transcriptional data set to infer regulatory networks corresponding to these neuronal subtypes. For this purpose, we applied a constrained fuzzy logic (cFL) approach[2] to systematically compare gene interaction relationships within a proposed regulatory network against our neuronal *in vivo* transcriptional data set. Because the measured transcriptional profiles were in response to an overall physiological perturbation, as opposed to direct neuronal stimulation or perturbation, we have modified the original cFL approach to consider graded gene expression values, as observed *in vivo*, that stimulate or perturb the gene regulatory network. Subsequently we trained a proposed *a priori* network composed of causal gene interactions (obtained from literature) involved in the angiotensin II type 1 receptor-mediated pathway, a target of several hypertensive treatments. Using this approach we investigated the regulatory networks contributing to the observed distribution of neuronal phenotypes.

Our analysis revealed distinct regulatory network model structures corresponding to neuronal subtypes, distinguished by transcription factor regulation and feedback inhibition of neuronal functional genes. Using these distinct network models, we examined how a distribution of neuronal subtypes may arise by simulating individual network models over a range of gene expression level inputs and evaluating their *in silico* transcriptional outputs using multidimensional scaling. *In silico* network simulations produced a range of neuronal subtypes similar to what was experimentally observed and showed that distinct gene expression level inputs to distinct networks yield similar transcriptional phenotypes despite their differing regulatory relationships. This stands in contrast to the previous expectation of distinct networks producing distinct transcriptional phenotypes given identical inputs. These results suggest that regulatory network structure and network response to inputs contribute to a range of neuronal subtypes. Consequently, we can now use our models to explore how changes to regulatory network structures may alter the balance of neuronal subtypes within a phenotypic population and contribute to the dysregulation of blood pressure as observed in hypertension.

1. Park J, Brureau A, Kernan K, Starks A, Gulati S, Ogunnaike B, Schwaber J, Vadigepalli R. 2014. Inputs drive cell phenotype variability. Genome Res.
2. M. K. Morris, J. Saez-Rodriguez, D. C. Clarke, P. K. Sorger, and D. a Lauffenburger, "Training signaling pathway maps to biochemical data with constrained fuzzy logic: quantitative analysis of liver cell responses to inflammatory stimuli.," PLoS Comput. Biol., vol. 7, no. 3, p. e1001099, Mar. 2011.

Research Support: NIH (NIGMS R01 GM083108; NHLBI R01HL111621; NIAAA 5 T32 AA007463-26)

389953 Spatial Confinement and Protein Diffusion Modulate Stochastic Switching in Bistable Signaling Networks with Positive Feedback

Abstract:

Many key biochemical reactions that mediate signal transduction in cells occur at the cell membrane, yet how the two-dimensional membrane environment influences the collective behavior of signaling networks is poorly understood. To gain insight into the role of the membrane in signal transduction, we use computational methods that explicitly account for stochastic fluctuations to study signaling networks in which a protein is activated by means of a positive feedback loop. Positive feedback is a common motif in cell signaling networks that can allow cells to make binary decisions in response to graded stimuli. This study is motivated by activation of the small GTPase Ras, which is central to T cell activation in response to antigen. Focusing on a bistable regime, we investigate the effects of spatial confinement and protein diffusion on stochastic switching between inactive and active states. Independently varying the geometry of the system and protein diffusion coefficients allows tuning of the *in silico* system from one that mimics the cytoplasmic environment to one that mimics the membrane environment. We find that confining the system to a membrane-like environment markedly alters the emergent behavior and that spatially localized concentration fluctuations can result in the formation and growth of regions with high concentrations of active proteins. The process by which small clusters of active proteins nucleate the growth of active regions is the primary mechanism of stochastic switching from inactive to active states and is promoted by slow diffusion. We use analytical methods to investigate the chemical master equation governing the system dynamics to gain additional insight into the likelihood of stochastic switching in the bistable regime.

382762 Systems Analysis of Signal Transduction in Insulin Mediated PI3K/AKT Pathway in Self-Renewal State of Human Embryonic Stem Cells

Abstract:

Motivation: Human Embryonic Stem Cells (hESCs) are an attractive raw material for regenerative medicine due to their unique properties of self-renewal and lineage specific differentiation. While the therapeutic potential of the differentiated lineages is well known, production of large quantities of the desired lineage requires abundant supply of pure starting material. During self-renewal, hESCs proliferate to give more copies of them and this is initiated by inhibition of differentiation signals and maintenance of signals that promote proliferation. As a result, self-renewal provides an avenue for obtaining large quantities of the raw material. Like any biological process, self-renewal is also prone to variability and rational methods are necessary to control and maintain a robust state. Currently, this is achieved by experimental means alone. While experiments have identified the key signaling players, knowledge of the systems level functioning of the signaling pathways in maintaining a robust self-renewal state is still under-developed. In this work, we have analyzed the systems level signal transfer properties of an important self-renewal pathway using a mechanistic mathematical model adapted for hESC specific behavior. Our primary goal is to identify robustness promoting mechanisms in the pathway using a quantitative framework in conjunction with efficient computational tools.

Methods: The phosphorylation dynamics of the key components of the insulin mediated PI3K/AKT pathway was first analyzed by experimentally stimulating H1 hESCs with 100 nM insulin after growth factor starvation. The dynamics was then compared to *in silico* simulations using a detailed mechanistic differential equation model of insulin mediated PI3K/AKT pathway (20 state variables, 25 parameters in the current form) by Sedaghat *et al.* [1]. Global sensitivity analysis (GSA) was used to identify the key interactions in the pathway that affect the dynamics of self-renewal molecules like p-AKT. We utilized a computationally efficient algorithm called Random Sampling High Dimensional Model Representation (RS-HDMR) to capture sensitivity under combinatorial interactions between the model parameters. We recently validated the application of this algorithm for a signal transduction model [2]. In this work, using GSA, we identified robustness promoting mechanisms that ensure (1) maintenance of a first order or overshoot dynamics of self-renewal molecule, p-AKT and (2) robust transfer of signals from oscillatory insulin stimulus to p-AKT in the presence of noise. For the latter, we tested various oscillatory scenarios of insulin stimulus to study the noise filtering capabilities of the PI3K/AKT pathway. The robustness metrics propounded by Kitano were used to quantify the input-output relationships [3].

Results and Discussion: In the first theme, we analyzed the dynamics and steady states of four major nodes of the PI3K/AKT pathway. These include the active insulin receptors (p-IR), tyrosine IRS1 (p-IRS1 (Y)), serine IRS1 (p-IRS1 (S)) and p-AKT. Dynamics from insulin stimulation experiments showed that intracellular components of PI3K/AKT in hESCs show the typical overshoot behavior influenced by negative feedback by p-IRS1 (S). Using phosphatase PTP as a control point to modulate the dynamics, we found that presence of negative feedback acts as a robustness promoting mechanism. For hESC specific parameter ranges that promote overshoot behavior, the system is insensitive to perturbations associated with the cascade reactions that propagate signals from the active receptors to the downstream kinases (or the direct trunk of the pathway topology). We predict that if the system looses its negative feedback character or shifts to a regime where PTP levels are very low, it will become vulnerable to the otherwise insensitive perturbations. In such a regime, the levels of self-renewal molecules like p-AKT will be high but also highly variable. This sets up a classical robustness tradeoff often seen in complex systems.

In the second theme, we evaluated the efficiency of signal transfer for time dependent variation in the input signal. Our results indicate that oscillations of small frequency (ω) and high amplitude (α) are transduced down the pathway, but with amplitude attenuation under nominal conditions. Our results also demonstrate that the downstream molecules follow the main signal with very high fidelity even in the presence of noise. Any modulation of upstream positive regulator IRS1 (p-IRS1 (Y)) can result in amplification or attenuation of the signals. Increasing the oscillation frequency, however, results in a regime where significant attenuation of the amplitude may be achieved. Finally, at very high frequencies ($\log_{10}(\omega \text{ min}^{-1}) > 0.5$), all oscillations are cut-off. This region was not affected by parameter perturbations. This shows that the PI3K/AKT pathway has an intrinsic minimal response time and when the oscillations are faster than this response time, they are not transmitted.

Conclusions: Our detailed mechanistic model and experimental analysis identified the precise mechanisms to modulate self-renewal molecules like p-AKT. Our results show that faithful transfer of signal from the stimulating ligand to p-AKT occurs even in the presence of noise, albeit with signal attenuation and high frequency cut-off. This ensures that informative signals are transmitted down the pathway and high frequency noise is cut-off. Negative feedback contributes to signal attenuation, while positive regulators upstream of p-AKT contribute to signal amplification. Quantitative measures of robustness can be used to finely tune the signal transfer process in hESCs to ensure that the level as well as the variability is kept within sufficient limits. Further, understanding the parameter dependences of the signal filtering and cut-off processes can help in the design of optimal input stimulation scenarios to modulate the pathway. Our work presents a framework towards the design of targeted growth media to maintain robust cellular fate of hESCs.

References:

[1] Sedaghat *et al.* A mathematical model of metabolic insulin signaling pathways. American J of Physiol Endocrinol and Metabol (2002) 283: E1084-E1101

[2] Mathew *et al.* Bioinformatics 2014, Article in Press, ID: bioinf-2013-2200.r1 (btu209).

[3] Kitano, H. (2007). Towards a theory of biological robustness. Mol Syst Biol, 3, 137.

364726 A Neural Network Based Human Platelet Calcium Calculator Trained By Pairwise Agonist Scanning

Abstract:

Platelet activation occurs through multiple signaling pathways in which agonists bind specific receptors on the platelet to trigger signaling in a dose-dependent manner. During a clotting episode, platelets respond to exposed surface collagen, released ADP and synthesized thromboxane, and the protease thrombin, all while being simultaneously modulated by endothelial derived nitric oxide and prostacyclin. These receptor-mediated signaling pathways are not independent and significant crosstalk occurs between these signaling pathways. During blood clotting, numerous agonists bind surface and intracellular platelet receptors to regulate intracellular calcium mobilization, $[Ca(t)]_i$. Since platelet $[Ca(t)]_i$ controls granule release, COX-1 and integrin activation, shape change, and phosphatidylserine exposure, the accurate prediction of clotting requires prediction of platelet $[Ca(t)]_i$in response to multiple agonists.

Pairwise Agonist Scanning (PAS) deployed all single and pairwise combinations of 6 agonists (ADP, convulxin, thrombin, U46619, iloprost and GSNO used at 0.1, 1, and 1x EC50; 154 conditions) to stimulate platelet P_2Y_1/$P2Y_{12}$, GPVI, PAR-1/PAR4, TP, IP receptors, and guanylate cyclase, respectively, in apixaban-treated dilute platelet rich plasma (Fluo-4 loaded for $[Ca(t)]_i$ measurement).[1]This assay allows the contribution of SOCE and includes the pathophysiologic signaling distal of thrombin. Apixaban-treatment of the platelet rich plasma also enables the use of exogenously added thrombin as an agonist in this plasma-based assay. Iloprost and GSNO were also included in this assay to recapitulate endothelial-derived prostacyclin and nitric oxide effects on platelet function. PAS on 10 healthy donors (50 % male) in duplicate allowed training of 10 neural networks (NN, 2-layer/12-nodes). Additionally, trinary stimulations were conducted for the donors at all 0.1x and 1x EC50 doses (160 conditions).
The NN accurately predicted $[Ca(t)]_i$ for binary stimulation ($R = 0.994$) as expected for successful NN training, but also accurately predicted the traces of $[Ca]_i(t)$ for trinary stimulation ($R = 0.916$), a prediction beyond the PAS training set. The trinary synergy score, a normalized metric of signaling crosstalk based upon integrated $[Ca(t)]_i$, was also well predicted ($R = 0.829$) for the 160 trinary conditions. Furthermore, the NN was able to predict with reasonable accuracy ($R = 0.78436$), $[Ca(t)]_i$ of all sequential additions of pairs of ADP, convulxin, and U46619 at low medium and high doses (54 conditions).

For computations of clotting, a user can impose time-dependent combinations of the agonists at varying concentrations to estimate $[Ca(t)]_i$responses that are predictive of those found in healthy human blood. The 10-donor average NN represents a computationally fast and data-trained calcium calculator for platelet responses to 6 critical pathways that regulate coagulation.

References:

1. Chatterjee, M. S., Purvis, J. E., Brass, L. F., & Diamond, S. L. (2010). Pairwise agonist scanning predicts cellular signaling responses to combinatorial stimuli. (Nature Biotechnology).

384407 Predicting Gene Expression Data from Changes in Eicosanoid Metabolite Levels in Raw 264.7 Macrophages Using Cybernetic Variables

Abstract:

Since their inception nearly three decades ago, cybernetic models have been used to analyze the dynamic control of metabolism based off of goals relevant to an organism's functioning. In the case of single cellular organisms, goals like the maximization of carbon uptake rate or growth rate have shown much success in predicting phenomena like dynamic intracellular fluxes, gene knockout behavior and hysteresis effects in chemostat cultures. The goal of this work is to take this existing framework for modeling single-celled organisms and modify it appropriately to describe more complex mammalian metabolic networks. A key step in this process is finding an appropriate description of the metabolic objectives of these cells or cellular subsystems.

The focus of study in this work is the response of eicosanoid metabolism in RAW 264.7 macrophages to lipopolysaccharide (LPS), a biochemical marker of bacterial infection. Eicosanoids are a diverse group of molecules derived from the oxidation of fatty acids that play a role in wide range of functions including signaling, inflammation and immunity. For this system, a cybernetic model was developed to describe the change in eicosanoid levels as a response to LPS and also in a control scenario with no LPS. The objective function used for this system is the maximization of the rate of TNF-alpha production which is a key player in inflammation. After fitting this model to metabolite data, the model's cybernetic variables were compared with dynamic gene expression data at the major branch points in this network. The comparison shows a significant level of agreement. This result serves as an important validation of cybernetic variables using real data from cells. Moreover, it shows that cybernetic variables can be used to infer trends in gene expression data only using information taken from the metabolite level and a description of the metabolic network's organizing principle.

388358 Analysis of Hormone and Mitogenic Signal Integration in Prostate Cancer

Abstract:

Prostate cancer is the most common cancer in men and the second leading cause of cancer- related death in the United States. Androgens, such as testosterone, are required for prostate cancer growth. Androgen ablation in combination with radiation or chemotherapy remains the primary non-surgical treatment for androgen-dependent prostate cancer. However, androgen ablation typically fails to permanently arrest cancer progression. As testosterone is withdrawn, malfunctioning cells lose their androgen sensitivity and proliferate without hormone. These testosterone insensitive cells often result in Androgen-Independent Prostate Cancer (AIPC). AIPC is closely related to metastasis and decreased survival. In this study, we developed a physiochemical model of an archetypal network, namely, hormone and growth factor signaling in human prostate cancer cells. The model describes the integration of two simultaneous extracellular signaling inputs, namely, androgen and growth factors into a G1/S cell-cycle checkpoint decision (including the regulation of Cyclin D alternative splicing). The model, containing 780 species and 1674 interactions, was identified using data generated from 43 studies in LNCaP cell lines. The model focuses on the outlaw pathway, in which constitutively or overly-activated receptor tyrosine kinases (RTKs) stimulate Akt and mitogen-activated protein kinases (MAPKs), which in turn activate cytosolic androgen receptor (AR) in the absence of androgen. The model was validated with multiple experiments in LNCaP cell lines. In addition data from drug trials was used to evaluate the model's performance predicting the response of prostate tumors to specific drugs. For example, in response to Enzalutamide, an AR inhibitor, the model predicted a decline in PSA expression similar to patient response. Sensitivity analysis, conducted over an ensemble of prostate signaling models, suggested that in an androgen free environment general translation was more sensitive in androgen dependent cells, while in androgen independent cells the PI3K/MAPK pathway species were more sensitive. In a constant DHT environment sensitive species were conserved between the cell lines. These results suggest targeting the PI3K/MAPK pathway in addition to anti androgen therapies as a treatment for AIPC. Experiments using androgen dependent (LNCaP) and androgen independent (C4-2) cell lines in the presence of BEZ235, Enzalutamide, and Sorafenib were conducted to validate our model's findings.

Anti-proliferative effects and metabolic dynamics of menaquinone (Vitamin K2) supplementation to human "acute T cell leukemia" cells

Suganthagunthalam Dhakshinamoorthy, Jeffrey Skolnick, Mark Styczynski

Abstract

CoMet, a computational metabolomics tool, was designed to predict changes in metabolite levels between normal cells and cancerous cells based only on transcriptional data and database information such as metabolic pathways and enzyme function annotations. These predictions are further extended with the hypothesis that metabolites predicted to be depleted in cancerous cells may have anti-proliferative effects. Based only on this data, CoMet previously predicted that menaquinone (vitamin K2) would show anti-proliferative activity, which has previously been demonstrated. Here, we will discuss our experimental validation of CoMet's predictions based on flow cytometry and systems-level analysis of hundreds of metabolite concentrations in human acute T cell Leukemia (Jurkat cells) and normal lymphoblast cells.

We performed toxicity assays with C12-resazurin and sytox green dyes, analyzed the number of live and dead cells with flow cytometry to show that menaquinone has selective, concentration- and time-dependent toxicity effects on Jurkat cells. Up to 75% of cancerous cells were killed by menaquinone at 30μM, with only a minimal anti-proliferative effect observed in normal lymphoblasts at the same concentration. This provided finer resolution of the effects of menaquinone on Jurkat cells than the previously known results.

These anti-proliferative effects are likely to have a significant impact on the metabolic state of these cells, based at least in part on the fact that menaquinone's antiproliferative effect was predicted solely based on transcription of metabolic genes. Little is known about the changes in cancer cell metabolism caused by menaquinone supplementation. To this end, intracellular and extracellular metabolic samples were analyzed using 2D gas chromatography coupled to mass spectrometry for both Jurkat cells and lymphoblasts in time series experiments with different concentrations of menaquinone supplementation. Multivariate statistical analysis was used to analyze the metabolomic changes in the cancer cells and normal cells at baseline and due to menaquinone supplementation.

Keywords: metabolomics, cancer, leukemia, menaquinone, vitamin K2,

389948 Metabolic Shifts Toward Glutamine Regulate Tumor Growth and Invasion in Ovarian Cancer

Abstract:

Glutamine can play a critical role in cellular growth in multiple cancers. Glutamine addicted cancer cells are dependent on glutamine for viability, and their metabolism is reprogrammed for glutamine utilization through the tricarboxylic acid (TCA) cycle. Here, we have uncovered a missing link between cancer invasiveness and glutamine dependence. Using isotope tracer flux analysis, we found that low invasive ovarian cancer (OVCA) cells are glutamine independent, whereas high invasive OVCA cells are markedly glutamine dependent. Consistent with our findings, OVCA patients' microarray data suggest that glutaminolysis correlates with poor survival. Notably, the ratio of gene expression associated with glutamine anabolism versus catabolism has emerged as a novel biomarker for patient prognosis. Significantly, we found that glutamine regulates the activation of STAT3, a mediator of signaling pathways which regulates cancer hallmarks in invasive OVCA cells. Our findings suggest that a combined approach of targeting high invasive OVCA cells by blocking glutamine's entry into the TCA cycle, along with targeting low invasive OVCA cells by inhibiting glutamine synthesis and STAT3 may lead to potential therapeutic approaches for treating OVCAs.

367941 Metabolic Switching of Gluconeogenesis in Hepatoma Cells

Abstract:

Diabetes is a metabolic disease characterized by a high level of blood glucose caused by the abnormal regulation of glucose production in the liver. The study of the metabolic regulation of gluconeogenesis in liver can correspondingly provide us valuable information for drug evaluation and diagnosis. Here, we studied gluconeogenic metabolism using ^{13}C-metabolic flux analysis (^{13}C-MFA), a powerful methodology capable of revealing *in vivo* metabolic activity. ^{13}C-MFA methodology can quantify reversible, cycling and branched reactions in biochemical networks by the introduction of isotopic tracers, e.g. ^{13}C or ^{2}H atom-based substrates, and mass spectrometry. As a model system for gluconeogenesis, we used a rat hepatoma cell line that can produce glucose in media containing gluconeogenic precursors, e.g. glycerol, lactate and glutamine. We activated gluconeogenesis in hepatoma cells using a cAMP analogue as a phosphoenolpyruvate carboxykinase (PEPCK) transcriptional regulator and by dexamethasone, a glucocorticoid. Based on the culture conditions, we selected isotopic tracers that allowed us to validate the gluconeogenesis model and to estimate central metabolism as well as gluconeogenesis metabolic flux: [U-^{13}C]glycerol to quantify upper gluconeogenesis and the pentose phosphate pathway, [U-^{13}C]glutamine for the TCA cycle and [U-^{13}C]lactate for the pathway validation of gluconeogenesis. Moreover, we constructed a metabolic model based on ^{13}C-labeling information of intracellular metabolites from the multiple tracer experiments. The metabolic model covers gluconeogenesis, glycolysis, the pentose phosphate pathway, pyruvate and TCA cycles, amino acid metabolism, ketone body and albumin production, and fatty acid metabolism. Furthermore, we performed combined ^{13}C-MFA which enables us to observe whole metabolic information by the simultaneous estimation of labeling data from multiple tracer experiments, rather than fragmentized results. Surprisingly, we observed metabolic switching of gluconeogenesis by the cAMP analogue and glucocorticoid. The cAMP analogue showed PEPCK activation by increasing glucose production, but dexamethasone bypassed PEPCK flux to pyruvate kinase as well as glucose production. Dexamethasone shows bifunctional roles in mediating anabolism and catabolism during gluconeogenesis through controlling pyruvate cycling. This study provides a cell-based platform to evaluate metabolic disease and drug efficacy.

390109 Metabolomic Analysis of Neural Differentiation in Mouse Embryonic Stem Cells

Abstract:

Pluripotent stems cells are able to differentiate into any cell type of the three germ layers, and thus hold much promise for clinical therapy. Therefore, improvements in the fledgling field of stem cell biomanufacturing, or the production of stem cells *in vitro* on a commercial scale, are of critical importance to realizing stem cells' medical potential. However, despite extensive research, differentiation to desired cell types *in vitro* is non-trivial to assess and control. Recently, the metabolic remodeling that occurs during differentiation has been gathering considerable attention as not only an indicator of differentiation, but also as a potential means to regulate it. The systems-level dynamics of metabolism in differentiation are still largely unexplored, but will be vital to a deeper understanding of differentiation that would inform applications of metabolic control.

To achieve this goal, we have characterized the metabolic changes during differentiation of mouse embryonic stem cells using untargeted gas chromatography-mass spectrometry metabolomics, which provides semi-quantitative data for a wide range of highly diverse metabolites. As proof of principle, we first measured metabolic profiles of both intracellular and extracellular samples of mouse embryonic stem cells during differentiation down a neural lineage. While the progress of differentiation was initially verified by previously-established protein and gene expression markers using RT-PCR and flow cytometry, we were able to detect distinct intracellular metabolic changes even at very early times during the process of differentiation. The impacts of medium renewal on metabolism were investigated, as were the impacts of culture format on the measured metabolite profiles. Samples over the course of differentiation clustered into a number of distinct groups based on their sampling time and the degree of differentiation of the cell population. Thus, monitoring metabolic change over the time course of differentiation was shown to offer both another way to describe the transient cellular identities found in heterogeneously differentiating populations and insight into how this transition might be better controlled.

Palmitate-induced alterations in phospholipid composition promote ER stress and cellular dysfunction in hepatic lipotoxicity

Alexandra K. Leamy[1], Robert A. Egnatchik[1], Masakazu Shiota[2], Pavlina T. Ivanova[3], David S. Myers[3], H. Alex Brown[3,4,5], Jamey D. Young[1,2,]

[1]Department of Chemical and Biomolecular Engineering

[2]Department of Molecular Physiology and Biophysics

[3]Department of Pharmacology, Vanderbilt University Medical Center

[4]Department of Biochemistry

[5]Vanderbilt Institute of Chemical Biology

Lipotoxicity induced by saturated fatty acid (SFA) accumulation has been postulated as a contributing factor in the pathogenesis of diseases such as non-alcoholic fatty liver disease (NAFLD) and type II diabetes. Our central hypothesis is that overexposure to SFAs results in increased saturation of ER membrane phospholipids, which induces ER stress marked by calcium efflux and CHOP overexpression, ultimately culminating in apoptosis. Our objective is to determine how perturbations of pathways involved in lipid metabolism modulate SFA-induced lipotoxicity, as assessed by measurements of phospholipid (PL) composition, ER stress, mitochondrial dysfunction, and apoptotic cell death. Interestingly, treatment of either primary hepatocytes or H4IIEC3 cells with the SFA palmitate (PA) resulted in dramatic dilation of the ER membrane, accompanied by other markers of organelle dysfunction. PA treatment also caused increased de novo glycerolipid synthesis, marked by significant elevation of dipalmitoyl phosphatidic acid, diacylglycerol, and total phospholipid content in H4IIEC3 cells. Co-supplementation with the monounsaturated fatty acid oleate (OA) diverted PA flux into triacylglycerols, resulting in elevated lipid droplet accumulation but significantly reversing PA-induced increases in PL saturation and all other phenotypic markers of hepatic lipotoxicity. Our data indicate that modulating alternative pathways of lipid disposal can significantly reduce PL saturation and thereby suppress the lipotoxic effects of SFAs on hepatic cells.

This work is funded by National Science Foundation Career Award 0955251. Masakazu Shiota is supported by NIH DK060667. Partial support for the work is also provided by National Institutes of Health Grant U54 GM69338 (to HAB).

Glucocorticoids are steroid hormones produced by the adrenal cortex that have diverse effects on a variety of physiological processes including carbohydrate, lipid and protein metabolism, immune-regulation, bone homeostasis and developmental processes (1, 2). The properties of glucocorticoids in regulating the immune system are exploited clinically with numerous synthetic corticosteroids (such as dexamethasone, prednisolone, fluticasone and methylprednisolone) that are used as anti-inflammatory and immunosuppressive agents in the treatment of a variety of conditions (2, 3). However, because of their strong effects on systemic metabolism, long-term usage of corticosteroids can cause many side-effects including metabolic syndrome, dyslipidemia, steroid-induced diabetes, atherosclerosis and muscle atrophy (4, 5).

Through their diverse physiological effects, corticosteroids can influence the functioning of many tissues; liver being one of their primary targets. As the master regulator of the systemic energy balance; liver plays a central role in maintaining glucose and lipid homeostasis in the body (6-8). Corticosteroids have been shown to affect almost all functions in liver through multiple mechanisms of action; and long-term use of corticosteroids results in dysregulation of these processes causing diverse side-effects (5). Immense diversity of glucocorticoid action in liver is due to the diversity of molecular mechanisms as corticosteroids can interfere with the natural processes of ligand-receptor binding, nuclear translocation and regulation of transcription through both enhancing and inhibiting gene expression (9, 10). In addition, studies have shown that corticosteroids can regulate pathways by signaling through its receptor in a transcription-independent manner, although the exact mechanisms for the non-genomic effects are still unclear (10).

This complexity of molecular mechanisms necessitates a systems approach for identifying the impacts of corticosteroids on the different functions of liver (11). Previous studies on profiling the temporal changes in mRNA expression in multiple tissues following corticosteroid administration in adrenalectomized rats characterized the global dynamics of the system that are regulated at the transcriptional level (12-15). Although this information is useful and highly relevant, direct profiling of the protein expression changes and integrating the information from the genomic and proteomic response have potential to provide much deeper insights into corticosteroid actions, given their complex mechanisms. Recently, we profiled the hepatic proteome following methylprednisolone (MPL) administration in adrenalectomized rats through a novel high-throughput methodology based on ion-current LC/MS strategy (16). Availability of these rich time-series datasets for changes in both mRNA and protein expression allowed integration of both transcriptional and translational profiles, enabling a unique analysis to compare and contrast the two. The present study compiles complementary hepatic transcriptional and translational profiles induced by the exposure to MPL at a pharmacological dose, uncovering the interesting relationships between the temporal expression of mRNA and proteins that provides a basis for understanding the effects of the drug actions at different physiological levels.

Hierarchical clustering of the common genes represented in both mRNA and protein datasets displayed two dominant patterns. First cluster was populated with genes, for which corresponding mRNA and protein expression profiles were essentially parallel in direction; while for the second cluster, this directionality was reversed. Genes coding for heat shock proteins and proteins functioning in the regulation of protein degradation and translation machinery were among the first cluster with complementary mRNA and protein profiles. Temporal expression changes in mRNA that directly translated to concurrent or slightly delayed protein expression changes in this cluster indicated that these processes might be controlled at the transcriptional level and can represent direct transcriptional effects of CS. In contrast, genes in the second cluster, which are associated with the modulation of oxidative stress, lipid metabolism and bile-acid biosynthesis, appear to be regulated at post-transcriptional levels since the transcriptional profiles are not emulated by their protein counterparts. This suggested that in addition to the direct transcriptional effects of CS, there could be additional translational or post-translational effects, about which current knowledge is very limited.

References

1. Vegiopoulos A, Herzig S. Glucocorticoids, metabolism and metabolic diseases. Molecular and cellular endocrinology. 2007;275(1-2):43-61.
2. Barnes PJ. Anti-inflammatory actions of glucocorticoids: molecular mechanisms. Clin Sci (Lond). 1998;94(6):557-72.
3. Swartz SL, Dluhy RG. Corticosteroids: clinical pharmacology and therapeutic use. Drugs. 1978;16(3):238-55.
4. Bialas MC, Routledge PA. Adverse effects of corticosteroids. Adverse drug reactions and toxicological reviews. 1998;17(4):227-35.
5. Schacke H, Docke WD, Asadullah K. Mechanisms involved in the side effects of glucocorticoids. Pharmacology & therapeutics. 2002;96(1):23-43.
6. Morand EF, Leech M. Glucocorticoid regulation of inflammation: the plot thickens. Inflammation research : official journal of the European Histamine Research Society [et al]. 1999;48(11):557-60.
7. Andrews RC, Walker BR. Glucocorticoids and insulin resistance: old hormones, new targets. Clin Sci (Lond). 1999;96(5):513-23.
8. Hazra A, Pyszczynski NA, DuBois DC, Almon RR, Jusko WJ. Modeling of corticosteroid effects on hepatic low-density lipoprotein receptors and plasma lipid dynamics in rats. Pharm Res. 2008;25(4):769-80.

9. Oakley RH, Cidlowski JA. Cellular processing of the glucocorticoid receptor gene and protein: new mechanisms for generating tissue-specific actions of glucocorticoids. The Journal of biological chemistry. 2011;286(5):3177-84. PMCID: 3030321.
10. Schaaf MJ, Cidlowski JA. Molecular mechanisms of glucocorticoid action and resistance. The Journal of steroid biochemistry and molecular biology. 2002;83(1-5):37-48.
11. Nguyen TT, Almon RR, Dubois DC, Jusko WJ, Androulakis IP. Comparative analysis of acute and chronic corticosteroid pharmacogenomic effects in rat liver: transcriptional dynamics and regulatory structures. BMC bioinformatics. 2010;11:515. PMCID: 2973961.
12. Almon RR, DuBois DC, Jusko WJ. A microarray analysis of the temporal response of liver to methylprednisolone: a comparative analysis of two dosing regimens. Endocrinology. 2007;148(5):2209-25.
13. Almon RR, DuBois DC, Pearson KE, Stephan DA, Jusko WJ. Gene arrays and temporal patterns of drug response: corticosteroid effects on rat liver. Functional & integrative genomics. 2003;3(4):171-9.
14. Almon RR, DuBois DC, Yao Z, Hoffman EP, Ghimbovschi S, Jusko WJ. Microarray analysis of the temporal response of skeletal muscle to methylprednisolone: comparative analysis of two dosing regimens. Physiological genomics. 2007;30(3):282-99.
15. Almon RR, Lai W, DuBois DC, Jusko WJ. Corticosteroid-regulated genes in rat kidney: mining time series array data. American journal of physiology Endocrinology and metabolism. 2005;289(5):E870-82. PMCID: 3752664.
16. Tu C, Li J, Bu Y, Hangauer D, Qu J. An ion-current-based, comprehensive and reproducible proteomic strategy for comparative characterization of the cellular responses to novel anti-cancer agents in a prostate cell model. Journal of proteomics. 2012;77:187-201.

Chronic liver disease and cirrhosis were the twelfth leading cause of death in the US, accounting for ~30,000 victims. While orthotopic liver transplantation is a viable treatment option, approximately only 6,000 of 17,000 candidates on the waiting list receive an organ, partially due to the stringent criteria used to determine graft quality. The criteria are based on donor characteristics, such as prolonged warm ischemia times (WIT), donor age, and degree of steatosis, but the cut-offs are fairly arbitrary and many marginal livers that could have been transplanted are unfortunately also discarded. To expand the current donor pool, *ex vivo* machine perfusion of livers prior to transplantation has been proposed as a platform to metabolically recondition suboptimal grafts. The process of machine perfusion is also conducive to obtaining more quantitative markers for liver function to determine whether or not the organ is suitable for transplantation. Recently, the use of subnormothermic machine perfusion (SNMP) on rat livers with large WIT has been shown to improve post-transplantation survival [1]. However, the underlying mechanism by which machine perfusion alleviates ischemic injury, and the corresponding metabolic biomarkers that correlate with improved function have not been extensively examined.

In this study, we characterize the dynamics of 20 discarded human livers (deemed unsuitable for transplantation) during SNMP by performing a targeted metabolomics analysis on time-course biopsies, collected every hour for 3 hours of perfusion. Approximately 150 intracellular metabolites from primary metabolism and 300 lipid compounds are quantified and mapped to a metabolic network of hepatocyte metabolism. Briefly, we find that TCA cycle intermediates have elevated levels at later time-points during perfusion. Moreover, perfusion appears to improve intracellular ATP levels as well as other redox measures, namely NADPH/NADP and NADH/NAD. Interestingly, reduced glutathione (GSH) significantly drops during perfusion relative to oxidized glutathione (GSSG), suggesting free radical formation requiring GSH scavenging. We find that the levels of some intracellular metabolites to be strongly correlated to donor characteristics, such as ornithine for WIT and GSH/GSSG for degree of steatosis. In conjunction with metabolic flux data obtained from measuring extracellular metabolite levels in the media, these intracellular metabolomics data provide insight into the dynamics of liver perfusion to determine which metabolic pathways may be up or down-regulated. The results open doors to enable rational design strategies to improve organ recovery during perfusion, and provide novel quantitative markers to assess liver health prior to transplantation.

1. Berendsen TA, Bruinsma BG, Lee J, D'Andrea V, Liu Q, et al. (2012) A simplified subnormothermic machine perfusion system restores ischemically damaged liver grafts in a rat model of orthotopic liver transplantation. Transplantation research 1: 6. doi:10.1186/2047-1440-1-6.

390357 Serum Metabolomics of Bottlenose Dolphins

Abstract:

Metabolomics is a powerful downstream method to study the response of organisms to perturbations they may encounter either naturally or due to external factors. This technique is being increasingly brought to bear on different model organisms to explore their metabolic repertoire and get a better understanding of their biochemical workings. One such organism is the bottlenose dolphin, *Tursiops truncatus*. This is an extremely important animal studied for a variety of reasons, not the least of which being that they are apex predators in the food chain and thus likely reflective of overall ecosystem health. Of late there has been increased concern about the conservation status of these animals, as they are subject to numerous sources of natural and anthropogenic stresses. The physiological impact of these stresses is unclear, including the potential for anthropogenic stressors compounding the negative effects of otherwise tolerated natural stressors.

Here we present our work investigating the metabolic differences between wild and managed care dolphins to assess the impacts of different potential stress environments on their physiological state. Serum samples were acquired, in many cases from existing studies, that allowed the assessment of the metabolic differences between four different populations of dolphins using two-dimensional gas chromatography coupled to mass spectrometry. These groups are believed to have different levels of anthropogenic stress based on the level of handling and interaction they have with humans. Numerous differences were detected between these four groups which do not seem to be accounted for through variety of diet alone. Moreover, we also found that while chemicals annotated as metabolites are sufficient to provide some separation between the different classes, inclusion of unannotated metabolites via untargeted metabolomics significantly improves the separation between the groups. This suggests that the differences between these animals may be driven by portions of metabolism not typically focused upon in metabolomic studies, and may drive future efforts to characterize these analytes.

368573 Modeling Blood Flow Control in the Kidney

Abstract:

The ability of vascular beds to maintain nearly constant blood flow despite large fluctuations in blood pressure is known as blood flow autoregulation. Impaired autoregulation in the kidney is both a symptom of and a contributing factor to the progression of diseases such as hypertension and diabetes. Obtaining a better understanding of renal autoregulation under both physiological and pathological conditions will lead to improved methods for controlling the progression of kidney-related diseases. Mathematical modeling based on experimental observations has become an important tool for investigating and understanding mechanisms of autoregulation and feedback in the kidney and is used here to describe blood flow control in the kidney under healthy and diabetic conditions. The model presented here couples two autoregulation mechanisms to the transport of chloride ions in the kidney, generating new insight into how blood flow in the kidneys of healthy and diabetic patients changes in response to increased blood pressure or impaired regulatory mechanisms.

The mathematical model developed in this study describes the dynamics of blood flow regulation in a single nephron, the functional unit of the kidney. The model includes two primary mechanisms of kidney autoregulation, the myogenic response and tubuloglomerular feedback (TGF), which constrict or dilate the afferent arteriole (vessel that supplies blood to the nephron upstream of the glomerular filtration unit) in response to changes in blood pressure and chloride ion concentration, respectively. The model is composed of two parts: (1) a millimeter-scale blood vessel wall mechanics model describing the nonlinear effects of the myogenic and TGF responses and (2) a molecular-scale mechanistic model of chloride ion transport in the loop of Henle (the salt/fluid reabsorption portion of the nephron downstream of the glomerular filtration unit). The model also accounts for the lag time dynamics associated with communicating the measured chloride concentration at the outlet of the loop of Henle to the smooth muscle cells of the afferent arteriole. Model parameters for physiological and diabetic conditions were taken or estimated from the experimental literature. The resulting system of delay differential equations was solved numerically using the delay differential equation solver *DDE23* in MATLAB.

The model was used to study the isolated and combined effects of the myogenic and TGF mechanisms as blood pressure varies. The model predicts stable glomerular filtration rate within a physiological range of perfusion blood pressure (60–180 mmHg). The results indicate that the myogenic response contributes more significantly than the TGF mechanism to the autoregulation of blood flow, and the contribution of TGF to overall autoregulation is significant only within a narrow band of perfusion pressure values near the baseline. The coupled dynamic model predictions of kidney blood flow are consistent with experimental observations [1] [5]. Model simulations of renal autoregulation in diabetic patients show a significant 60% increase in kidney blood flow, largely due to simulated impairment of voltage-gated calcium ion channels of the smooth muscle cells in the afferent arteriole that prevent constriction of the vessel. The present model was also used for bifurcation analysis to investigate how changes in key model parameters and system perturbations affect autoregulation dynamics.

References:

[1] LG Navar et al., Kidney Int, 1998, 54:S17-21.

[2] RJ Roman, Am J Physiol, 1986, 251:F57-65.

[3] RJ Roman et al., Am J Physiol, 1985, 248:F199-205.

[4] CM Sorensen et al., Am J Physiol, 2000, 279:R1017-1024.

[5] SJ Van Dijk et al., Kidney Int, 2006, 69:1369-1376.

Emergent Properties of the *E. coli* Nitric Oxide Response Network

Abstract

Nitric oxide (NO•) is an antimicrobial weapon synthesized by the innate immune system to inactivate pathogens, and its potency has motivated recent efforts to design exogenous NO• delivery systems for treating and preventing infection. As such, the identification of bacterial targets that increase NO• sensitivity upon perturbation would prove highly valuable in designing novel therapeutics that synergize with host-derived or exogenously-administered NO•. However, the broad reactivity of NO• and its metabolites, as well as the systems-level response it elicits in microbes, necessitates a computational approach to comprehensively and quantitatively understand the system. We previously constructed and experimentally validated a kinetic model of the NO• biochemical reaction network in *Escherichia coli*, capable of predicting the distribution and fate of NO• delivered to a bacterial culture. Using the model, we investigated the control of NO• dynamics in *E. coli* cultures, and had found that the primary aerobic detoxification enzyme, NO• dioxygenase (Hmp), loses utility as the NO• delivery rate increases, and functions as a dominant NO• consumption pathway even at O_2 concentrations as low as 35 μM (microaerobic). Recently, the model was used to identify and characterize an emergent NO•-protective role for the Clp protease in *E. coli*. An *E. coli* Δ*clpP* mutant exhibited prolonged bacteriostasis following NO• treatment, and a reduced rate of NO• consumption compared to the wild-type. A parametric analysis was conducted to reconcile the model-predicted NO• dynamics in a Δ*clpP* culture with experimental measurements, and impaired Hmp transcription, translation, or catalytic activity were identified as potential sources of the phenotype. Pre-expression of Hmp in wild-type and Δ*clpP* cultures before NO• treatment demonstrated a negligible difference in catalytic activity, narrowing the search to Hmp transcription and/or translation. Quantification of protein and mRNA transcripts using an Hmp–GFP translational fusion and real-time quantitative PCR, respectively, established a transcription deficiency to be the primary cause of the NO•-sensitive Δ*clpP* phenotype. Interestingly, the transcription defect of NO•-stressed Δ*clpP* was observed to persist among two additional expression systems, each differing in promoter and protein sequence, suggesting a broad effect that is not specific to Hmp. This work demonstrates the utility of quantitative modeling in its ability to direct experimentation to decipher the underlying mechanisms of mutant phenotypes, establishing a more accurate understanding of clinically-relevant microbial stress responses, and ultimately aiding in the rational design of therapeutics.

377248 On Inverse Problems in Complex Biological Systems: An Integrated Multiple Shooting and Genetic Algorithm Approach

Abstract:

The inverse problem of parameter estimation and state identification from dynamical data obtained from systems exhibiting nonlinear behavior, using ODE model descriptions is crucial for studying the behavior of complex biological systems. The inverse problem is generally posed as optimization problem so as to fit a suitable model description by using monitored data from the system. Several deterministic, heuristic and metaheuristic methods have been used for these studies (Chou and Voit, 2009; Engl et. al., 2009). Employing deterministic approaches for parameter estimation in nonlinear systems often face the problem of getting stuck in local minima while heuristic/metaheuristic approaches become computationally expensive due to the high number of iterative simulations of the ODE model.

To alleviate some of these problems, hybrid optimization approaches have been proposed (El-Milhoub et. al., 2006; Rodriguez-Fernandez et. al., 2006). In the majority of these algorithms, heuristic global search approaches, such as genetic algorithms, simulated annealing etc. are used along with deterministic gradient based local search algorithms. In addition, several support methods have also been developed that narrow down the search space and ease the computational and mathematical rigor for the original optimization problem (Voit and Almeida, 2004; Kimura et. al., 2005; Jia et. al., 2011; Vilela et. al., 2009). Even so, these hybrid and support methods have limited applicability and there is still a need for the development of algorithms that address the common issues faced in finding solution of inverse problems. These pertain to presence of measurement noise, missing data points and when some process variables are not amenable to measurement.

Recently, we proposed a novel parameter estimation and state identification algorithm, EMSGA, by Embedding (E) a multiple shooting (MS) methodology in the framework of a genetic algorithm (GA) (Sarode et. al., 2014). In the MS component, the total sampling interval of the monitored data is divided into multiple sub-intervals for numerical solution and the data fitting objective is minimized such that consecutive sub-intervals have continuity (Peifer and Timmer, 2007, Ghosh et. al., 2001). This optimization problem can be solved using gradient search methods such as nonlinear constrained Sequential Quadratic Programming (SQP). On convergence it provides the state identification of process variables along with true process parameter values by estimation. Its application offers advantages in handling noisy data, missing data points and recovering unmonitored process variables. However, the application of MS as a stand-alone formalism for complex optimization problems with high nonlinearity, stiffness, large number of parameters, and noisy data is inefficient because of the weak global search capabilities. Thus, in our ESMGA approach, the MS methodology is now embedded in a framework of a heuristic global search method, viz., GA and combining the two alleviates the difficulties. In contrast to standard stand alone GA, where based on the objective score of estimated parameter values, a new generation population is evolved directly, in EMSGA, the MS algorithm is incorporated as a transformer function to obtain even better estimates of the guessed parameters and process variables values before evolving the estimated population by GA. Thus, the local search of MS and global search of GA work in cooperation to estimate process parameter values, noise free process state variables values while simultaneously recovering the unmonitored dynamics of the other dependent variables. To assist in the computational performance of the algorithm, we show how additional transformer functions in the form of different support methods can further augment the overall EMSGA framework in a modular fashion.

We demonstrate the utility of the proposed algorithm in modeling noisy data obtained from batch fermentation for ethanol and glycerol production using *Saccharomyces diastaticus* LORRE 316 (Wang et. al., 2001). Ability of estimating large number of parameters along with noise-free dynamics of monitored and unmonitored system variables is demonstrated using noisy oscillatory and even chaotic dynamics obtained from Lotka-Volterra models (Voit and Chou, 2010). Canonical models for fermentation pathway in *Saccharomyces cerevisiae* (Curto et. al., 1995) are used to demonstrate utility of various support methods in the performance enhancement of the algorithm.

Chou, I., & Voit, E. O. (2009). Recent developments in parameter estimation and structure identification of biochemical and genomic systems. *Mathematical Biosciences*, *219*(2), 57-83.

Curto, R., Sorribas, A., & Cascante, M. (1995). Comparative characterization of the fermentation pathway of *Saccharomyces cerevisiae* using biochemical systems theory and metabolic control analysis: Model definition and nomenclature. *Mathematical Biosciences*, *130*(1), 25-50.

El-Milhoub, T. A., Hopgood, A. A., Nolle, L., & Battersby, A. (2006). Hybrid Genetic Algorithms: A Review. *Engineering Letters*, *13*(3), 124-137.

Engl, H. W., Flamm, C., Kügler, P., Lu, J., Müller, S., & Schuster, P. (2009). Inverse problems in systems biology. *Inverse Problems*, 25(12), 123014.

Ghosh, A., Ravi Kumar, V., & Kulkarni, B. D. (2001). Parameter estimation in spatially extended systems: The Karhunen-Lóeve and Galerkin multiple shooting approach. *Physical Review E*, *64*(5), 056222_1-056222_8.

Irvine, D. H., & Savageau, M. A. (1990). Efficient solution of nonlinear ordinary differential equations expressed in S-system canonical form. *SIAM Journal on Numerical Analysis*, *27*(3), 704-735.

Jia, G., Stephanopoulos, G. N., & Gunawan, R. (2011). Parameter estimation of kinetic models from metabolic profiles: two-phase dynamic decoupling method. *Bioinformatics*, *27*(14), 1964-1970.

Kimura, S., Ide, K., Kashihara, A., Kano, M., Hatakeyama, M., Masui, R., Nakagawa, N., Yokoyama, S., Kuramitsu, S., & Konagaya, A. (2005). Inference of S-system models of genetic networks using a cooperative coevolutionary algorithm. *Bioinformatics*, 21, 1154-63.

Peifer, M., & Timmer, J. (2007). Parameter estimation in ordinary differential equations for biochemical processes using the method of multiple shooting. *Systems Biology, IET*, *1*(2), 78-88.

Rodriguez-Fernandez, M., Egea, J. A., & Banga, J. R. (2006). Novel metaheuristic for parameter estimation in nonlinear dynamic biological systems. *BMC Bioinformatics*, *7*(1), 483.

Sarode, K. D., Ravi Kumar, V., Kulkarni, B. D. (2014) Embedded Multiple Shooting Methodology in a Genetic Algorithm Framework for Parameter Estimation and State Identification of Complex Systems. *Industrial & Engineering Chemistry Research* Submitted.

Savageau, M. A., & Voit, E. O. (1987). Recasting nonlinear differential equations as S-systems: a canonical nonlinear form. *Mathematical Biosciences*, *87*(1), 83-115.

Vilela, M., Vinga, S., Maia, M. A., Voit, E. O., & Almeida, J. S. (2009). Identification of neutral biochemical network models from time series data. *BMC Systems Biology*, *3*(1), 47.

Voit, E. O. (2013). Biochemical systems theory: a review. *ISRN Biomathematics*, *2013*.

Voit, E. O., & Almeida, J. (2004). Decoupling dynamical systems for pathway identification from metabolic profiles. *Bioinformatics*, *20*(11), 1670-1681.

Voit, E., & Chou, I. C. (2010). Parameter estimation in canonical biological systems models. *International Journal of Systems and Synthetic Biology*, *1*, 1-19.

Wang, F. S., Su, T. L., & Jang, H. J. (2001). Hybrid differential evolution for problems of kinetic parameter estimation and dynamic optimization of an ethanol fermentation process. *Industrial & Engineering Chemistry Research*, *40*(13), 2876-2885.

377538 The Influence of Peristalsis Enhanced Permeability on Drug Delivery in the Gastrointestinal Tract

Abstract:

Here we evaluate the influence of peristalsis on delivery of pharmaceutical agents into the esophagus. Although drug delivery to the esophagus remains challenging (bolus passage completes in <20 s), it is increasingly important. For example, debilitating esophageal diseases including eosinophilic esophagitis (EoE) require delivery of small molecules such as the glucocorticoids fluticasone and budesonide to the lumenal surface of the esophageal. These molecules counter damage to the mucosal surface caused by the invasion of eosinophils, a type of white blood cell, into the esophagus. Yet, how esophageal peristalsis affects drug delivery remains unclear. For example, we know experimentally that as the bolus moves from the oral cavity to the stomach, the esophageal wall moves in the opposite direction (towards the oral cavity) introducing both a tensile strain on the esophageal wall and retrograde surface velocities. This talk will discuss each of these effects and their influence on the rate of small molecule transport through the mucus and tissue.

379770 Long-Term Memory in Bacterial Chemotaxis

Abstract:

The chemosensory system in *E. coli* is a paradigm for two-component signaling pathways. The receptors form the input and the flagellar motors form the output of the chemotaxis network. Receptor modification through methylation/demethylation provides a short-term memory (~ few seconds) that enables cells to make temporal comparisons of ligand concentrations, and hence to respond to chemical gradients during chemotaxis. Recent work has now identified a long-term memory (~ few minutes), enabled by stimuli-dependent structural changes in the assembly of the flagellar motor. I will discuss how we experimentally determine the output dose-response curves from the remodeling of single motors. I will then describe our theoretical model that provides a comprehensive description of the network input-output relationship at the single cell level.

A modular approach to mathematically modeling liquid-solute transport in Cystic Fibrosis airways: toward patient-specific treatment design and control

Matthew R Markovetz[1], Timothy E Corcoran[1,2,6], Landon W Locke[2], Stefanie Brown[2], Michael M Myerburg[2], Joseph M Pilewski[2], Robert S Parker[1,3,4,5]

1. Department of Chemical and Petroleum Engineering, Swanson School of Engineering 2. Pulmonary, Allergy, and Critical Care Medicine 3. McGowan Institute of Regenerative Medicine 4. Clinical Research, Investigation, and Systems Modeling of Acute Illness (CRISMA) Laboratories, Department of Critical Care Medicine 5. University of Pittsburgh Cancer Institute 6. Department of Bioengineering, Swanson School of Engineering, University of Pittsburgh, Pittsburgh, PA, USA

Cystic Fibrosis (CF) is an autosomal-recessive disorder caused by the dysfunction or absence of the Cl^- secreting cystic fibrosis transmembrane conductance regulator (CFTR) and subsequent dysregulation of Na^+ absorption via the epithelial sodium channel (ENaC). The respective malfunction and hyperactivity of these ion channels in the epithelium of patients with CF causes abnormal liquid hyperabsorption. The primary disease effects occur in the lung, where liquid hyperabsorption leads to mucociliary clearance failure, infection, and inflammation, which necessitates patient rescue via lung transplantation. Current therapies for CF lung disease target the hyperabsorptive phenotype using inhaled osmotic agents to reverse the osmotic driving force that leads to liquid hyperabsorption in the lungs.

Modeling the phenomena that underlie the regulation of airway surface liquid (ASL) volume provides critical insights into the mechanisms that play a major role in the pathophysiology of CF. We have developed a compartment-based, mathematical representation of liquid and solute transport in human bronchial epithelial (HBE) culture under thin-film culture conditions, using ordinary differential equations (ODEs). The three model compartments represent the dynamic volumes of the cells, the ASL, and the basolateral bath. Both *in vivo* and *in vitro*, ASL volumes are osmotically controlled via the directed transport of solutes, primarily Na^+ and Cl^-, into and out of the cell. We have included the concentration of those ions in all three compartments as model states. Additionally, the volume-regulated transport of Na^+ and Cl^- via ENaC and CFTR, respectively, between the ASL and the cell are given saturating terms in our model, and the basolateral transporters of these ions are also included. To elucidate the different contributions of trans- and para-cellular liquid transport, we include apical mass of Tc-99m DTPA, a radiopharmaceutical indicator of paracellular liquid transport *in vivo* and *in vitro*. Finally, we introduce apical and basolateral states for mannitol, a candidate therapeutic osmotic agent that impacts osmotic gradients but is not trafficked by cells.

Model simulations were compared to experimental results from HBE cell cultures obtained following 10 μL apical volume challenges. In these experiments, 10 μL of either isosmotic (300 mOsm) Ringers solution or hyperosmotic (600 mOsm) Ringers and mannitol were applied to the surface of non-CF or CF HBE cultures. Model simulations of the isosmotic challenge follow the liquid absorption dynamics of non-CF HBE cells with the CFTR term present (the non CF case) and display hyperabsorptive dynamics in the case where the CFTR term is set to zero and ENaC permeability is increased (the CF case). The model is also able to capture ASL volume swelling and cellular shrinking observed after a 10 μL hyperosmotic challenge to either non-CF or CF HBE cultures. Setting the cellular Na^+ efflux term to zero in our model demonstrates the cellular swelling and reduced absorption that occurs after a 10 μL isosmotic challenge and subsequent basolateral addition of ouabain, an inhibitor of Na+ transport though cells. These results underscore the physiological relevance of our model and its utility as a means of predicting potential therapeutic effects at the cellular scale.

Our mechanistic model of transport in the bronchial epithelium describes the dynamics of liquid and solute transport under multiple osmotic and therapeutic challenges. This serves as the next step toward incorporating a cell-scale model into our organism-scale compartmental model of transport in the human pulmonary system. Our goal, ultimately, is to use this combined model, in concert with clinically applicable therapies and systems engineering tools, to design patient-specific therapeutic dosing regimens having the ability to mitigate the hyperabsorptive phenotype present in patients with CF.

Thermodynamics based Model Analysis of Deletion Mutagenesis in Mitochondrial DNA

[1, 2] Lakshmanan LN, [3] Gruber J, [4] Halliwell B, and [1]Gunawan R.

[1]Institute for Chemical and Bioengineering, ETH Zurich, Zurich, Switzerland

[2]Department of Chemical and Biomolecular Engineering, National University of Singapore, Singapore

[3]Yale-NUS College, Department of Biochemistry, Neurobiology and Ageing Program, National University of Singapore, Singapore

[4]Department of Biochemistry, Neurobiology and Ageing Program, Centre for Life Sciences, National University of Singapore, Singapore

Mitochondria are primarily responsible for the synthesis of ATP through the process of oxidative phosphorylation (OXPHOS). Moreover, mitochondrial DNA (mtDNA) encodes a subset of proteins required for OXPHOS. Deletion mutations in mtDNA have been observed to accumulate in aged tissues from humans and animal models. These deletions have been suggested to play a causal role in the degeneration of skeletal muscle fibres and neurons associated with aging and multiple neuromuscular disorders [1-3].

Deletion mutations involve DNA strand breaks at two breakpoint positions and the loss of the intervening sequence. Sequence analyses near observed deletion breakpoints in the literature have indicated the importance of direct repeat (DR) and homologous sequences in the region flanking the breakpoints. Based on these observations, a number of hypotheses suggest that mtDNA deletions occur via DNA misalignments involving DR motifs or homologous sequences during mtDNA replication or repair [4-7]. However, the most relevant sequence size undergoing misalignments *in vivo* remains unknown [4, 6, 7]. Also a considerable fraction of observed mtDNA deletions do not contain any homologous sequence flanking their breakpoints. Such deletions have therefore been suggested to occur via non-homologous recombination. Currently the relative importance of the misalignment-dependent (MD) and –independent (MI) mechanisms in mtDNA deletions remains poorly understood.

The lack of precise knowledge of mtDNA deletion mutagenesis hinders the formulation of effective therapies against mtDNA deletion associated disorders.

The objectives of our study are (1) to estimate the relative contributions of MD and MI mutagenesis in experimentally identified mtDNA deletions, (2) to investigate the relevant length(s) of DNA sequence misalignments associated with mtDNA deletions, and (3) to investigate the presence of species- and clinical condition-specific mechanism of mtDNA deletion mutagenesis. For these purposes, we have developed a probabilistic mixture model based on DNA-DNA hybridization thermodynamics, the underlying principle governing the DNA misalignments.

The model was developed by considering the probability of finding a deletion with breakpoint positions at $(l,r)_i$, where l is the left breakpoint and r is the right breakpoint positions in mtDNA sequence. For MD mechanism, we also considered multiple mtDNA segment sizes that could be involved in the misalignments. The probability associated with the i-th mtDNA deletion with the breakpoint position $(l,r)_i$ was calculated as follows:

$$p((l,r)_i\,;\,\theta) = \theta_{MI} * p_{MI}((l,r)_i) \;+\; \sum_{j \in J} \theta_{MD,j} * p_{MD,j}((l,r)_i) \qquad (1)$$

where θ_{MI} denotes the fraction of deletions associated with MI mechanism, $\theta_{MD,j}$ denotes the fractions of deletions due to misalignments of mtDNA of size j bp in the set J, and $p_{MI}((l,r)_i)$ and $p_{MD,j}((l,r)_i)$ represent the probability of finding MI and MD deletions at the position $(l,r)_i$ respectively. We performed the calculations above by considering two distinct (non-overlapping) partitions of the major arc of mtDNA, i.e the left breakpoint region (LBR) and the right breakpoint region (RBR), containing the majority of the left and right breakpoints. MI deletions were assumed to occur randomly across the space spanned by LBR vs. RBR, and thus p_{MI} was set as the uniform distribution. On the other hand, the distribution of MD-related deletions was assumed to be governed by the thermal stability of the misaligned duplex. The calculation of the probability $p_{MD,j}$ was based on the partition functions of DNA-DNA hybridization involving duplexes formed between two size *j* bp segments, one from the LBR and the other from the RBR. Using literature-compiled deletion breakpoints, we estimated the fractions of θ_{MI} and $\theta_{MD,j}$'s by a constrained maximization of the following log-likelihood function:

$$\max \log L(\theta) = \sum_{i=1}^{N} \log p\left((l,r)_i ; \theta\right) \quad (2)$$

subject to $\theta_{MI} + \sum_{j \in J} \theta_{MD,j} = 1$, where *N* is the number of mtDNA deletions in a particular dataset.

We validated the predictions of θ_{MI} and $\theta_{MD,j}$'s using computer generated deletion breakpoints with known fractions of θ_{MI} and $\theta_{MD,j}$'s. We then used the model to analyze datasets, consisting of 569 deletion breakpoints compiled from the literature and spanning 8 different clinical conditions in human, rhesus monkey, mouse and rat. The results from the model analysis suggested that MD mechanism is the predominant mtDNA deletion formation in human aging, monkey aging, mouse aging and human myopathy. Meanwhile, MI mechanism is the dominant factor in rat aging, double strand break (DSB) mouse and human polymerase gamma (POLG) mutant datasets.

A key finding from the analysis of the mtDNA deletions was that in datasets with high MD fractions, the most relevant misalignments involve short sequences of length ≤ 6 bp (predominantly 3 – 4 bp). Moreover, recent experimental studies on mtDNA replication using electron microscopy have indicated that only a small fraction of mtDNA remains single stranded during mtDNA replication [8, 9]. Taken together, the findings above suggested that misalignments of long homologous sequence are the exception rather than the norm *in vivo*, which stands in contrast to the prevailing hypotheses. The analysis of aging associated datasets in different animals further showed that age-related mtDNA deletions in human, rhesus monkey and mouse form by similar mechanisms, again one that involves the misalignments of short mtDNA sequences. Meanwhile, homology-dependent and -independent recombination mechanisms have been suggested earlier to cause mtDNA deletions during the repair of mtDNA DSBs [5, 6, 10]. The large MI fractions in our analysis of mtDNA deletion datasets associated with DSBs (i.e. mouse DSB and human POLG) suggested that non-homologous recombination as the main source of mtDNA deletions *in vivo* during repair of DSBs.

1. Bender, A., et al., *High levels of mitochondrial DNA deletions in substantia nigra neurons in aging and Parkinson disease.* Nat Genet, 2006. **38**(5): p. 515-517.
2. McKenzie, D., et al., *Mitochondrial DNA deletion mutations: A causal role in sarcopenia.* European Journal of Biochemistry, 2002. **269**(8): p. 2010-2015.
3. Yamashita, S., et al., *Genotype and phenotype analyses in 136 patients with single large-scale mitochondrial DNA deletions.* Journal of Human Genetics, 2008. **53**(7): p. 598-606.
4. Guo, X., et al., *Repeats, longevity and the sources of mtDNA deletions: evidence from 'deletional spectra'.* Trends in Genetics, 2010. **26**(8): p. 340-3.
5. Krishnan, K.J., et al., *What causes mitochondrial DNA deletions in human cells?* Nat Genet, 2008. **40**(3): p. 275-9.
6. Sadikovic, B., et al., *Sequence homology at the breakpoint and clinical phenotype of mitochondrial DNA deletion syndromes.* PLoS ONE, 2010. **5**(12).
7. Shoffner, J.M., et al., *Spontaneous Kearns-Sayre/chronic external ophthalmoplegia plus syndrome associated with a mitochondrial DNA deletion: a slip-replication model and metabolic therapy.* Proc Natl Acad Sci U S A, 1989. **86**(20): p. 7952-6.
8. Holt, I.J. and A. Reyes, *Human mitochondrial DNA replication.* Cold Spring Harb Perspect Biol, 2012. **4**(12).
9. Yasukawa, T., et al., *Replication of vertebrate mitochondrial DNA entails transient ribonucleotide incorporation throughout the lagging strand.* EMBO J, 2006. **25**(22): p. 5358-71.
10. Chen, T., et al., *The generation of mitochondrial DNA large-scale deletions in human cells.* J Hum Genet, 2011. **56**(10): p. 689-694.

388821 Nequalsone: A Model-Based Clinical Decision Support Tool for Implementing Personalized Treatment Strategies

Abstract:

Cancer treatment has been a challenge for more than a century to clinicians and researchers alike. Every individual is unique, both genetically and phenotypically, and hence respond to the treatment differently. Dosing decisions based on individual patients' characteristics offer maximum efficacy without causing severe toxicities. While this has largely been empirical in the past, quantitative predictive methods are gaining traction in the recent times as a tool to predict the oncoming efficacy/toxicity and make dose adjustments accordingly. Quantitative tools, suitably empowered by systems theoretic approach, could serve as a decision-support mechanism to physicians. However, use of such sophisticated models and algorithms have not found its way into clinical practice as they demand extensive knowledge of mathematical, statistical and computational background and skills. Obviously, clinicians, already burdened with their clinical practice, are not in a position to make full use of these tools. In other words, there are no user-friendly software available for clinicians to assist them in their day-to-day decision-making with regard to dosage adjustments.

In this work, we present nEqualsOne, a one-of-its-kind, model-based standalone software, to quantitatively predict the clinical outcome for a specific patient and determine the optimal chemotherapeutic dose to achieve a desired clinical response. It is a point-and-click, interactive software where the practicing clinicians can process the available patient information through mathematical models and predict as well as optimize the dose for each individual patient based on a defined clinical target. Finally, the dosing decision is made by the clinicians taking into account some information that are not considered in the software. We demonstrate an application for an important chemotherapeutic drug known as 6-Mercaptopurine in tailoring treatment for leukemia patients.

AUTHOR INDEX

AUTHOR INDEX

AIChE
120 Wall Street, FL 23
New York, NY 10005-4020

ISBN 978-1-5108-1247-5

9 781510 812475

CURRAN ASSOCIATES INC.
proceedings.com